MICROWAVE
CIRCULATOR DESIGN

MICROWAVE
CIRCULATOR DESIGN

Douglas K. Linkhart

AH

Linkhart, Douglas K.
  Microwave Circulator Design/ Douglas K. Linkhart.

  p.   cm.
  Bibliography: p.
  Includes index.
  ISBN   0-89006-329-X:
  1. Circulators, Wave-Guide—Design and Construction.   I. Title
TK7871.65.L56   1989   621.381′331—dc20   89-6550
ISBN   0-89006-329-X   CIP

Copyright © 1989
ARTECH HOUSE, INC.
685 Canton Street
Norwood, MA 02062

International Standard Book Number: 0-89006-329-X
Library of Congress Catalog Card Number: 89-6550

  10   9   8   7   6   5   4   3   2   1

Contents

Preface	vii
Acknowledgments	viii
Chapter 1 THEORY OF OPERATION	1
1.1 Units, Conversions, and Symbols	1
1.2 The Physical Basis of Ferrimagnetism	4
1.3 Ferrimagnetic Resonance	9
1.4 Microwave Propagation in Ferrites	13
Chapter 2 CIRCULATOR SPECIFICATION	25
2.1 The Parameters	25
2.2 Junction Circulators	37
2.3 Lumped-Constant Circulators	41
2.4 Differential Phase Shift Circulators	42
2.5 Field-Displacement Isolators	44
2.6 Resonance Isolators	46
Chapter 3 APPLICATIONS OF CIRCULATORS	51
3.1 Load Isolation	51
3.2 Duplexing	52
3.3 Multiplexing	57
3.4 Parametric Amplifiers	59
3.5 Phase Shifting	63
Chapter 4 MATERIAL SELECTION	67
4.1 Ferrite Selection	67
4.2 Magnet Selection	77
4.3 Magnetic Compensating Material Selection	79
4.4 Dielectric Selection	79
4.5 Metals Selection	80
Chapter 5 ELECTRICAL DESIGN	83
5.1 Junction Circulators	83

5.2 Lumped-Constant Circulators   112
5.3 Differential Phase Shift Circulators   117
5.4 Resonance Isolators   121
5.5 Dummy Loads for Isolators   124
Chapter 6 MAGNETIC DESIGN   129
6.1 Magnet Sizing   129
6.2 Shielding   139
6.3 Temperature Compensation   140
6.4 Completing the Circuit   141
Chapter 7 MECHANICAL DESIGN   145
7.1 Coaxial Junction Circulators   145
7.2 Lumped-Constant Circulators   158
7.3 Waveguide Circulators   160
7.4 Resonance Isolators   162
Chapter 8 ASSEMBLY AND TESTING   163
8.1 Assembly Techniques   163
8.2 Finding the Operating Point   168
8.3 Taking Data   171
Chapter 9 TUNING   173
9.1 Magnetic Adjustment   173
9.2 Electrical Adjustment   176
List of Symbols   187
Index   191

Preface

The objective of this book is to present theory, information, and design procedures that will enable microwave engineers and technicians to design and build circulators successfully. We begin in Chapter 1 with a discussion of the various units used in the circulator design computations, then we cover the theory of operation, but only to a depth sufficient to form a foundation for the rest of the text.

Those who prefer not to read through the theory of operation can skip to Chapter 2, which describes how to specify a circulator. Typically, the specification process comes before design and construction, so this chapter is placed near the front of the book. The beginner can become involved in circulator specification before learning about microwave magnetics.

Chapter 3 is about applications of circulators. The information presented here may give the engineer new ideas about how to solve problems using circulators. Knowledge of circulator applications will also help the designer do a better job.

Material selection criteria are presented in Chapter 4. The correct selection of materials, particularly ferrites, is very important in circulator design.

Chapters 5, 6, and 7 cover electrical, magnetic, and mechanical design, respectively. The order in which design procedures are presented in this book is the order in which they should be applied in practice. Thus, the reader can begin the design process before he or she finishes the book.

The remainder of the book is devoted to assembly and testing (Chapter 8) and tuning (Chapter 9), the benchwork involved in making a circulator work.

Field-displacement isolators and resonance isolators are not circulators at all, but their design parallels circulator design in many respects, so we have included some information about them (see Section 2.5).

I hope that this book fills the need for comprehensive circulator design procedures.

Acknowledgments

I am indebted to many early workers in the microwave industry for their research efforts. Thanks are also due to the folks at MPD/Micon who have established a work environment which has perpetuated my interest in microwave circulators and other components.

I wish to thank Ralph Esagui and Andrew Owens for their helpful comments and encouragement during the preparation of this book. They have both listened to me talk about circulators for hours, never once showing a lack of enthusiasm.

Last, but not least, thanks go to my wife and son, who complained only a few times about the time that I devoted to this book.

Chapter 1
Theory of Operation

1.1 UNITS, CONVERSIONS, AND SYMBOLS

There are several systems of standard units in use today. These include the CGS (centimeter-gram-second), MKS (meter-kilogram-second), SI (International System), and English systems. The SI system is now the most widely used.

When a book is written using an accepted standard system of units, it is often difficult for an American engineer to use. This is because many of us measure lengths in inches and angles in degrees (and probably will for some time). Ferrite and magnet manufacturers use CGS units, so a book written entirely in SI units would require us to perform still more conversions.

Table 1.1 shows the units that are used in this book. In many of the equations presented, particularly those which are part of derivations, the units are unimportant. In other equations, any units may be used so long as we are consistent. If we stick to the units shown in the table, difficulty in calculation should not arise.

The units in Table 1.1 are mostly SI units, with the exceptions of those for magnetic field intensity and magnetic induction, which are CGS units. Data for ferrite and magnet materials are normally presented in CGS format. Thermal conductivity is also not an SI unit, because the W/in°C unit is much more practical. The length, area, and angle units can be changed to inches, square inches, and degrees in many equations. Care should be exercised so as not to mix units from different systems in the same equation.

Some useful conversion factors are listed in Table 1.2. It may be necessary to change units in some instances.

Table 1.1 Units Used in this Book

Quantity	Unit	System	Other Units
Angle	rad	SI	deg
Length	m	SI	in
Area	m^2	SI	in^2
Time	s	SI	
Frequency	Hz	SI	
Velocity	m/s	SI	
Mass	kg	SI	g
Force	N	SI	
Torque	Nm	SI	
Energy	J	SI	
Power	W	SI	
Temperature	°C	SI	
Thermal conductivity	W/in°C	US	
Electric charge	C	SI	
Voltage (EMF)	V	SI	
Permittivity	F/m	SI	
Electric current	A	SI	
Magnetic field intensity	Oe	CGS	
Magnetic induction	G	CGS	
Permeability	H/m	SI	
Capacitance	F	SI	
Inductance	H	SI	
Resistance	Ω	SI	
Conductance	S	SI	
Conductivity	S/m	SI	
Impedance	Ω	SI	
Admittance	S	SI	
Susceptance	S	SI	
Isolation	dB		
Insertion loss	dB		
Return loss	dB		

Abbreviations Used in this Table

Abbreviation	Unit	Abbreviation	Unit
A	ampere	J	joule
°C	degree Celsius	kg	kilogram
C	coulomb	m	meter
dB	decibel	N	newton
deg	degree	Oe	oersted
F	farad	rad	radian
g	gram	s	second
G	gauss	S	siemen
H	henry	SI	International System
Hz	hertz	W	watt
in	inch	Ω	ohm

Table 1.2 Conversion Factors

Quantity	To convert	Into	Multiply by
Angle	rad	degrees	5.730×10^1
Length	m	in	3.937×10^1
Area	m²	in²	1.550×10^3
Temperature	°F	°C	5.556×10^{-1} *
Thermal conductivity	g cal/s cm² °C/cm	W/in°C	1.06×10^1
Thermal conductivity	BTU/hr ft² °F/ft	W/in°C	4.4×10^{-2}
Magnetic field intensity	A/m	Oe	1.257×10^{-2}
Magnetic induction	T (tesla)	G	1.000×10^4
Magnetomotance	ampere turns	gilberts	1.257×10^0

* First subtract 32, then multiply by factor shown.

To the uninitiated, the symbols for the ferrite material magnetization and saturation magnetization can be confusing. In the MKS system, magnetic induction, magnetic field intensity, and magnetization are related by

$$B = \mu_0 H + \mu_0 M$$

The equivalent of this equation in the CGS system is somewhat different:

$$B = H + 4\pi M$$

Usually, the factor 4π is included with the magnetization so that $4\pi M$ in the CGS system is equivalent to $\mu_0 M$ in the MKS system. Thus, the ferrite magnetization in the CGS system is denoted $4\pi M$ and the saturation magnetization is $4\pi M_s$. The subscript simply means saturation.

The gyromagnetic ratio, γ, is presented in the text as a constant relating the ferrimagnetic resonance radian frequency and the dc magnetic field. This constant is given the value of 2.80 MHz/Oe. The frequency here (MHz) is clearly not a radian frequency. The value of the gyromagnetic ratio in radian frequency/Oersted units is 17.6 Mrad/s/Oe. The reason for giving the constant a value in MHz/Oe is that the factor of 2π in any frequencies in the same equation with the gyromagnetic ratio can be eliminated to simplify calculation. Therefore, radian frequencies (ω or ω_0) that appear in the same equation with γ are not really radian frequencies, but frequencies in MHz.

The permittivity symbol, ε, represents either absolute or relative permittivity depending on the context. In general, it represents relative permittivity in equations that are used in the circulator design process and absolute permittivity in derivations. A complete list of symbols appears at the end of the text.

1.2 THE PHYSICAL BASIS OF FERRIMAGNETISM

All microwave circulators contain ferrimagnetic materials; therefore, it is important that we understand the theory behind the magnetic properties of these materials in order to understand circulator operation.

Ferrimagnetic materials were known some 3,000 years ago in the form of lodestones (magnetites). The first artificial ferrimagnets, or ferrites, as we shall refer to them from now on, were made in 1909 by Hilpert [1]. At the time, the composition of the ferrites could not be controlled reliably, so it was not until some time later that practical ferrites were successfully produced.

During World War II, J. L. Snoek developed usable ferrites [2] at the Philips Gloeilampenfabriken research laboratories in The Netherlands. These ferrites were originally intended for use at low frequencies, but it was soon discovered that the materials had many possibilities for use at microwave frequencies. A different class of ferrites, the garnets, were first produced [2] in the 1950s by the French physicist Louis Néel and by workers at Bell Laboratories.

There are three main classes of ferrites: the spinels, the garnets, and the hexagonal ferrites. The names of the ferrite classes describe the crystal structures of the materials. The spinels have the same crystal structure as the mineral spinel, and the garnets have the structure of the mineral andradite (common garnet or black garnet). The hexagonal ferrites, which have hexagonal crystal structures, are used primarily for ceramic magnets and devices that operate at millimeter wavelengths. We shall only concern ourselves with the spinels and garnets in the following discussion.

The ferrites are ionic crystals. That is, the atoms in the crystal are bonded together by ionic bonds—the ions either lose or accept electrons when they become bonded. A lowering of potential energy occurs when ions of opposite charge approach each other, and it is such a lowering of energy that the atoms seek. Because a crystal structure also leads to a lowered energy level, the ferrite molecules do not commonly exist separately but become arranged in the crystal. A ferrite can be thought of as one giant molecule.

Single crystal ferrites can be produced for some special applications, but most ferrites are polycrystalline. A polycrystal has a nonuniform orientation of the crystal lattice.

Ions in the ferrite have magnetic moments that lead to paramagnetism, which is the ability of the ions to be acted upon individually, with no mutual interaction, by a magnetic field. The magnetic moment of an ion is equal to the sum of two different moments: one due to the electron spin, and the other due to the orbital motion of the electrons.

The magnetic moments of the ions can be calculated [3] from the quantum numbers, but there is no reason to do so here. Atoms that have only filled shells have no net magnetic moment, because the electron spins and the electron orbital effects cancel each other. Only when there are unpaired electron spins or a net orbital angular momentum can there be paramagnetism. The net magnetic moment due to electron spins depends on the number of spins in each direction. If all the spins are up, the net moment is equal to the sum of the moments due to each spin. If all the spins are down, the net moment would be determined in the same manner. If some of the spins are up and others down, the difference in the number of spins in each direction (equal quantities of spins in each direction would cause complete cancellation) would determine the net magnetic moment.

Hund's rule states that electrons entering a subshell containing more than one orbital will be spread out over the available orbitals with their spins in the same direction. Figure 1.1 illustrates the determination of the net moment from the spins.

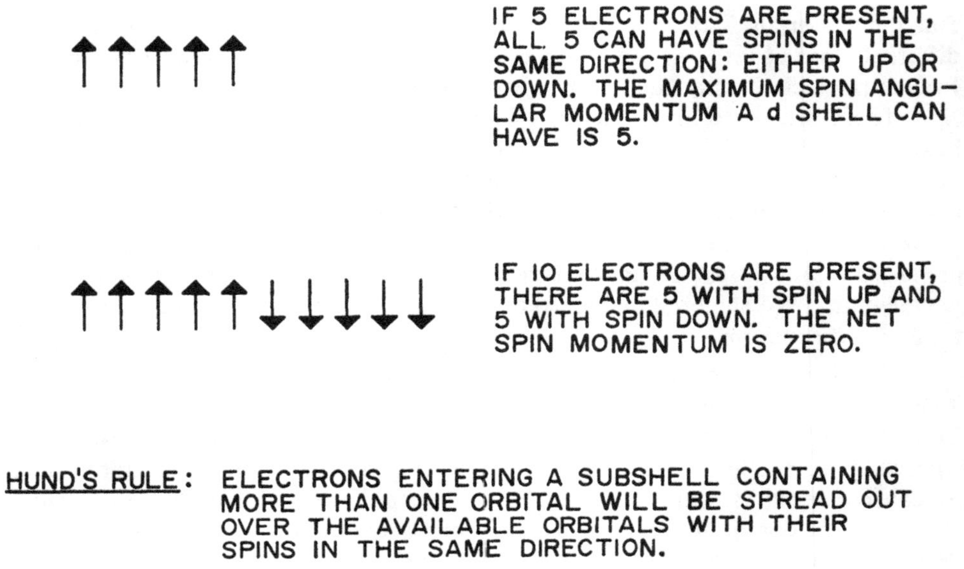

↑↑↑↑↑ IF 5 ELECTRONS ARE PRESENT, ALL 5 CAN HAVE SPINS IN THE SAME DIRECTION: EITHER UP OR DOWN. THE MAXIMUM SPIN ANGULAR MOMENTUM A d SHELL CAN HAVE IS 5.

↑↑↑↑↑↓↓↓↓↓ IF IO ELECTRONS ARE PRESENT, THERE ARE 5 WITH SPIN UP AND 5 WITH SPIN DOWN. THE NET SPIN MOMENTUM IS ZERO.

HUND'S RULE: ELECTRONS ENTERING A SUBSHELL CONTAINING MORE THAN ONE ORBITAL WILL BE SPREAD OUT OVER THE AVAILABLE ORBITALS WITH THEIR SPINS IN THE SAME DIRECTION.

Fig. 1.1 Determination of net spin angular momentum from electron spins.

The mineral spinel has the formula $(MgAl_2O_4)_8$. The microwave spinels (ferrites) have a similar general formula, $(MOFe_2O_3)_8$, where M represents a divalent metal such as iron, manganese, magnesium, nickel, zinc, cadmium, cobalt, copper, or a mixture of these. The metallic ions are small in comparison to the oxygen ions, so the crystal structure is determined primarily by the locations of the oxygen ions. The metallic ions fit in between the oxygen ions.

The ions are arranged in a *face-centered cubic* (FCC) lattice. Figure 1.2 shows the orientation of the ions if viewed in the direction of a body diagonal of the spinel crystal ((111) crystal plane).

There are two possible sites for the metallic ions in this FCC lattice [4]. These sites are shown in Figure 1.2. The A sites have as the four nearest neighbors the oxygen ions. These sites are said to have tetrahedral coordination because of the shape formed by the oxygen ions. Six nearest neighbor oxygen ions surround the metallic ions at the B sites. These sites have octahedral coordination.

There are also two types of octants in the spinel structure, shown in Figures 1.3 and 1.4. The spinel crystal would be made up of four of each of these octants. In Figure 1.4, three of the oxygen ions shown with the octahedral site are actually part of adjacent octants. The unit crystal contains 56 atoms: 32 oxygen atoms and 24 metallic atoms. Each unit cell also contains 8 tetrahedral sites and 16 octahedral sites. Note that the oxygen ions are shared by more than one site; the tetrahe-

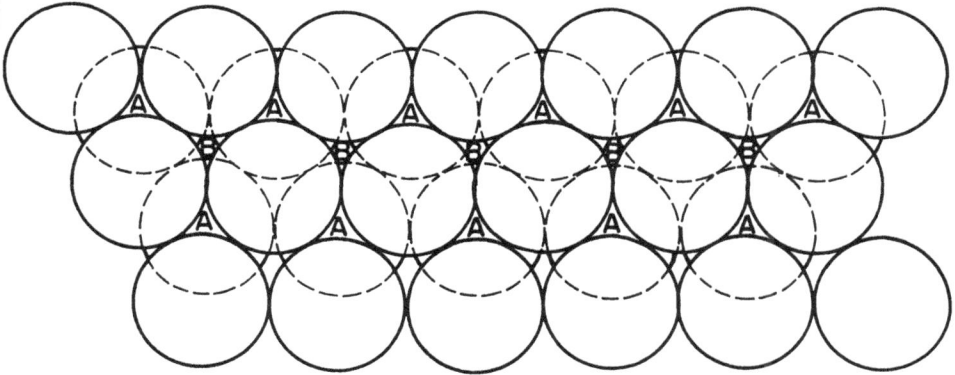

Fig. 1.2 Hexagonal close-packed (FCC) lattice: A = tetrahedral sites; B = octahedral sites.

drons and octahedrons are not really isolated as shown in Figures 1.3 and 1.4. Metallic ions with valences of $+3$ occupy all of the tetrahedral sites and half of the octahedral sites. The remaining octahedral sites are filled by $+2$ valence metallic ions.

The ferrimagnetic properties of the spinels result from the orientation of the net magnetic moments of the metallic ions. The moments of the ions at the tetrahedral sites cancel the moments of 8 of the ions at octahedral sites. The remaining 8 octahedral ions give the ferrite a net magnetic moment.

Ferrimagnetism should not be confused with ferromagnetism. Ferromagnetism is the spontaneous alignment of all the atomic magnetic moments. Antiferromagnetism is the complete cancellation of all the magnetic moments in the solid. Ferrimagnetism, then, is a hybrid of ferromagnetism and antiferromagnetism—there is an incomplete cancellation of the atomic magnetic moments.

Garnets, which have the general formula $5Fe_2O_33M_2O_3$, have crystal structures that are different from those of the spinels in two respects. First, in the garnet structure there are three types of sites for metallic ions, instead of the two site types in the spinel structure. Second, all the sites in the garnet structure are filled.

The M in the garnet formula represents yttrium or some other rare earth. Some of the iron in the garnet can be replaced by aluminum to vary the net magnetic moment.

A very important electrical property of ferrites is low conductivity. The molecules of ferrites can be engineered so that there are no mechanisms of conduction. A complete absence of conduction electrons in the crystal structure will ensure that the ferrite material has high resistance.

High resistance is important in microwave applications of ferrites because it

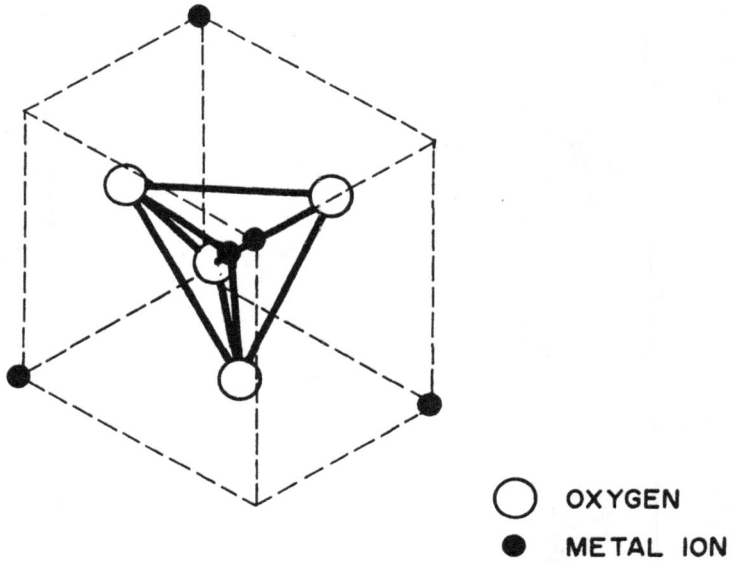

OXYGEN
METAL ION

Fig. 1.3 One octant of spinel crystal structure including tetrahedral site.

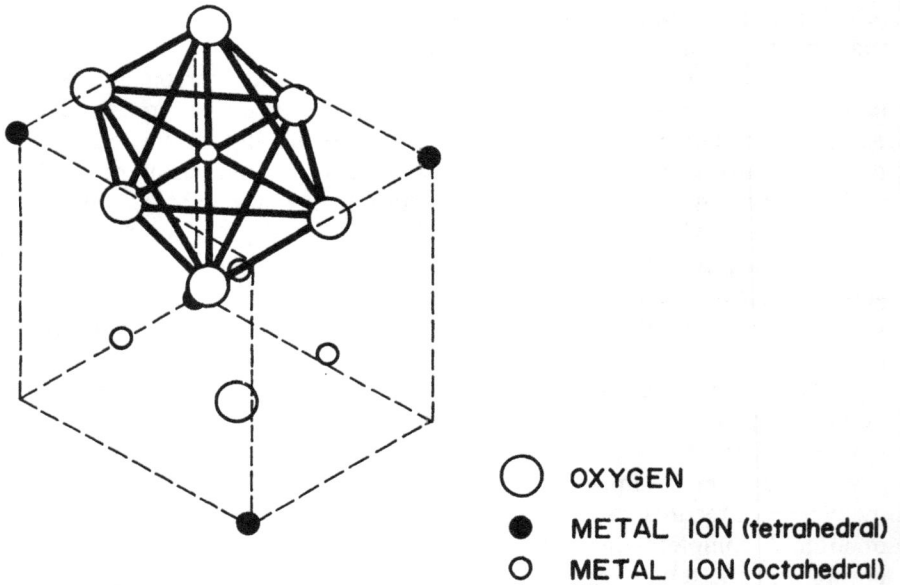

OXYGEN
METAL ION (tetrahedral)
METAL ION (octahedral)

Fig. 1.4 One octant of spinel crystal structure including octahedral site.

reduces eddy current losses. Before there were ferrites, there were iron powder materials. The key to reducing the eddy current losses in these materials·is to reduce the size of the iron particles and disperse them in an insulating material. The smallest possible particle of iron is an iron atom. Therefore, the development of ferrites was based on forming materials from iron atoms using chemical techniques.

1.3 FERRIMAGNETIC RESONANCE

We can more easily comprehend the concept of ferrimagnetic resonance if we first consider a bicycle wheel. If we hold a bicycle wheel by one end of the axle and spin the wheel, as shown in Figure 1.5, the axle will remain in a plane parallel to the ground and twist in the direction indicated. This gyroscopic effect can be explained mathematically with reference to the vectors in Figure 1.6(a).

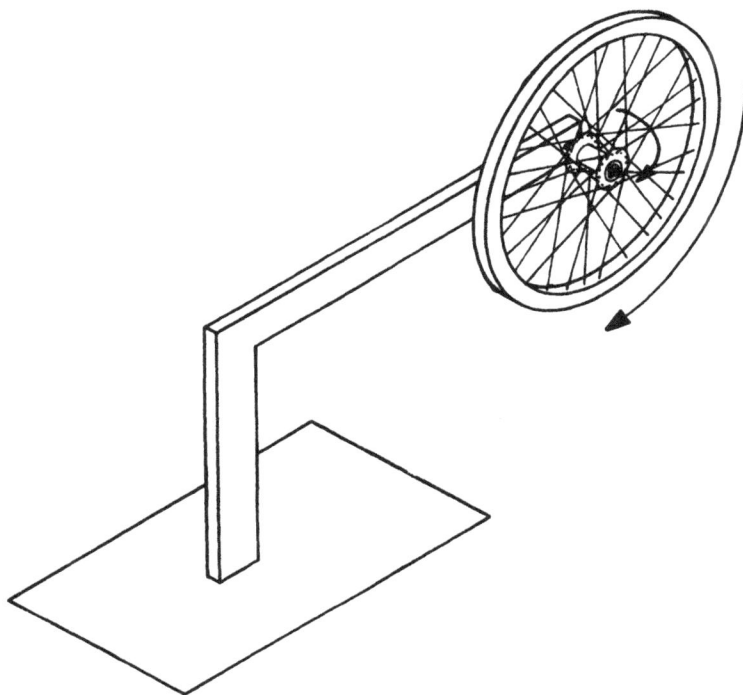

Fig. 1.5 Precession of a bicycle wheel.

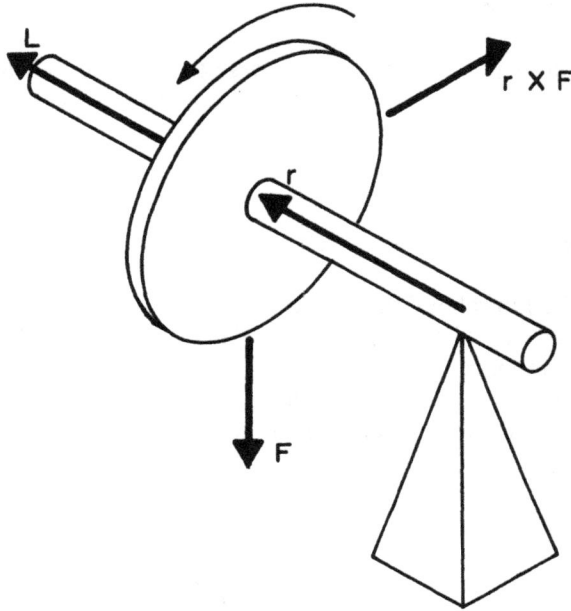

Fig. 1.6(a) Vector analysis of a gyroscope.

Because the angular momentum of this gyroscope system must be conserved [5], we have

$$\frac{d\mathbf{L}}{dt} = \mathbf{N} \tag{1.1}$$

where $\mathbf{L}$ is the total angular momentum and $\mathbf{N}$ is the total external torque applied. The torque applied to the system in Figure 1.6a by gravity is $(\mathbf{r} \times \mathbf{F})$, and is directed perpendicular to $\mathbf{L}$ and $\mathbf{r}$. Insofar as $d\mathbf{L}/dt$ is in the same direction as the torque, by Equation (1.1), $\mathbf{L}$ will precess in the direction of $(\mathbf{r} \times \mathbf{F})$. Note that the new direction of $\mathbf{L}$ after the infinitesimal time increment dt is simply

$$\mathbf{L} \text{ (new)} = \mathbf{L} \text{ (old)} + \frac{d\mathbf{L}}{dt} \tag{1.2}$$

We now depart from bicycle mechanics by replacing our bicycle wheel with a spinning electron in the ferrite material, depicted in Figure 1.6(b).

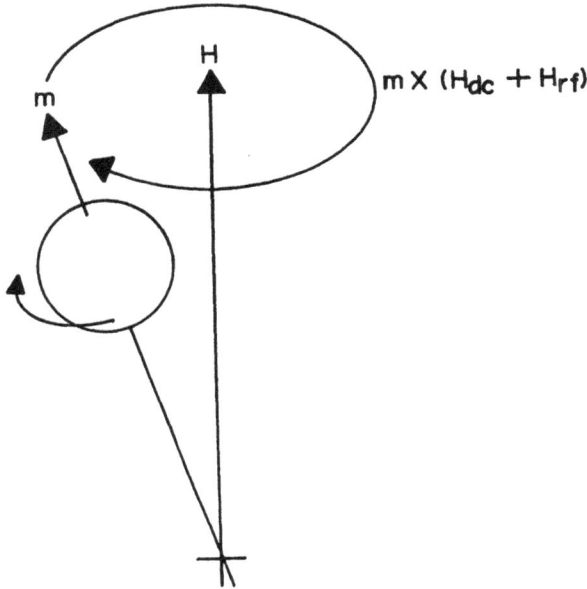

Fig. 1.6(b) Spinning electron in the presence of a magnetic field.

The spinning orbital electrons of the ions in a ferrite material, which have net magnetic moments (**m**), can be acted upon by a magnetic field in much the same way that the bicycle wheel is acted upon by gravity. The torque acting on the electron is given by

$$\mathbf{N} = \mathbf{m} \times (H_{dc} + H_{RF}) \tag{1.3}$$

where H_{dc} is an applied direct magnetic field, H_{RF} is a radio-frequency magnetic field applied to the ferrite, and **m** is the net magnetic moment of the electron.

The torques acting on the electrons cause them to precess or wobble. If no RF magnetic field is applied to the ferrite, the electron spin axes will become aligned with the dc magnetic field in a short time. This is because there is no longer any torque acting on the electron when its spin axis is aligned with the magnetic field.

When an RF magnetic field is applied to the ferrite in addition to a dc field, the electrons precess about the direction of the dc field. This precession was shown [6] by Landau and Lifshitz in 1935.

If we substitute the torque given by Equation (1.3) for $\mathbf{N}$ in Equation (1.1), and use $d\mathbf{m}/(g\,dt)$ as the first derivative of the electronic momentum, we have

$$\frac{d\mathbf{m}}{g\,dt} = \mathbf{m} \times (H_{dc} + H_{RF}) \tag{1.4}$$

where g is a constant—the Landé splitting factor or g factor, as it has come to be known. Ideally, the g factor is equal to 2, but because of orbital interactions (the electrons are not really independent particles), this factor differs slightly from 2. The exact value of the g factor depends on the specific ferrite material.

Because the electrons precess in a circular fashion, $d\mathbf{m}/dt$ can be replaced by $\mathbf{m}\omega$, where ω is the radian frequency of the precession. We can replace $\mathbf{m}$ with $-eh/4\pi m_0 c$ and $\mathbf{m}/g$ with $-h/4\pi$ to derive

$$\omega = \frac{e(H_{dc} + H_{RF})}{m_0 c} \tag{1.5}$$

where e is the unit electron charge, m_0 is the mass of the electron, and c is the velocity of light; $\mathbf{m} = -eh/4\pi m_0 c$ is the Bohr magneton, the natural unit of magnetic moment; $\mathbf{m}/g$ is the angular momentum. The electron spin number is $-\frac{1}{2}$, so the angular momentum is $-\frac{1}{2}h/2\pi = -h/4\pi$.

We now have an expression for the frequency with which the electrons precess about the dc magnetic field. If the frequency of the applied RF magnetic field is equal to the precessional frequency, the direction of the electron spin magnetic moment will depart greatly from the direction of the dc field. This is the ferrimagnetic resonance effect, discovered in 1946 by Griffiths. A theory of ferrimagnetic resonance, or gyromagnetic resonance as it is also called, was developed by Kittel in 1948.

At resonance, the only torque acting on the electrons is that due to the dc magnetic field, because the electrons are precessing at the frequency of the RF magnetic field. Therefore, at resonance, Equation (1.5) becomes

$$\omega_0 = \frac{eH_{dc}}{m_0 c} \tag{1.6}$$

where ω_0 is the radian frequency of the ferrimagnetic resonance. In MKS units, the factor $e/m_0 c$ in Equation (1.6) becomes $\mu_0 e/m_0$ where μ_0 is the permeability of free space ($4\pi \times 10^{-7}\ H/m$). This factor is called the gyromagnetic ratio, γ. We can rewrite Equation (1.6) as

$$\omega_0 = \gamma H_{dc} \tag{1.7}$$

The gyromagnetic ratio has the value of 2.80 MHz/Oe, and is useful for determining the ferrimagnetic resonance frequency if the dc magnetic field is known, or for determining the magnetic field required to cause resonance at a specific frequency.

It is important to note that the dc magnetic field here is the field acting on the electrons, which may differ from the magnetic field applied external to the ferrite material because of the ferrite sample demagnetization factors. These factors vary depending on the shape of the ferrite.

In the preceding discussion we have turned our attention to the effect of the magnetic fields on a single electron, but all the electrons in the ferrite sample in these magnetic fields precess in the same manner. Insofar as the RF magnetic field (and the dc field also, perhaps) may vary at different positions in the ferrite sample, we have made the assumption that the ferrite is small compared to a wavelength so that the magnetic fields in the ferrite are uniform.

The ferrimagnetic resonance is of great importance in the theory of microwave circulators. In some types of circulators, the resonance is used directly to perform the circulation (or isolation) function, while in other types of circulators, the magnetic operating point of the circulator is chosen to be near the ferrimagnetic resonance in order to take advantage of the nonreciprocal properties of the ferrites in the vicinity of resonance.

1.4 MICROWAVE PROPAGATION IN FERRITES

Michael Faraday discovered [7] in 1845 that a piece of glass becomes optically active when placed in a strong magnetic field. The Faraday effect occurs as shown in Figure 1.7. Plane-polarized light incident upon the glass in a direction parallel to the applied magnetic field becomes polarized in a different plane. The glass rotates the plane of polarization of the light. The amount of rotation is directly proportional to the applied magnetic field and to the distance the light travels through the glass.

A similar effect, also called Faraday rotation, occurs when microwaves of a specific polarization are incident upon ferrite material that is subjected to a magnetic field parallel to the direction of propagation of the microwaves. One of the first microwave circulators was the Faraday rotation circulator. This circulator is illustrated in Figure 1.8. The arrows indicate the direction of the electric vectors in the waveguides. The center section of the circulator is a circular waveguide, and a transition to a rectangular waveguide is located at each end. The rectangular guide at port 2 is rotated through 45 degrees with respect to the guide at port 1. Two other rectangular waveguides emerge radially from the circular waveguide at ports 3 and 4. The axes of these two guides are parallel to the electric vectors in the

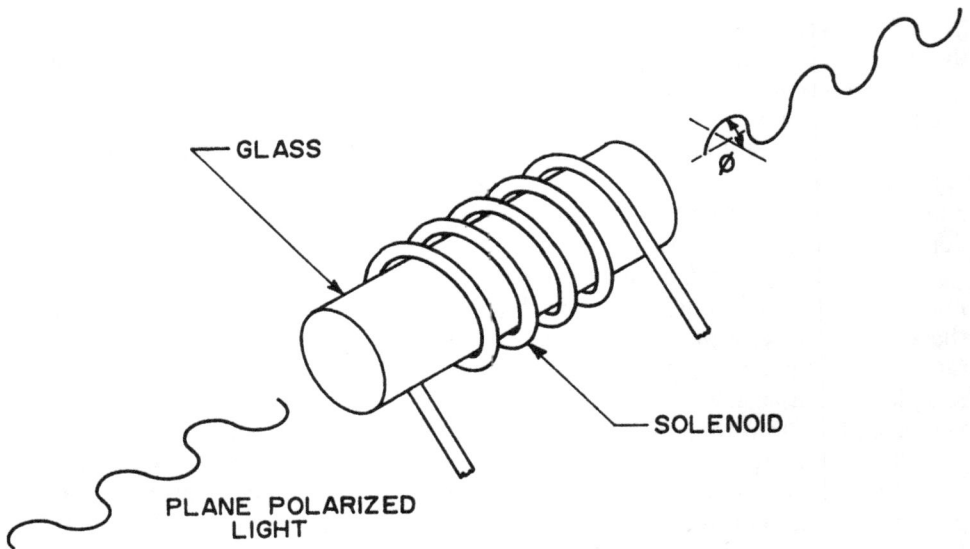

Fig. 1.7 The Faraday effect.

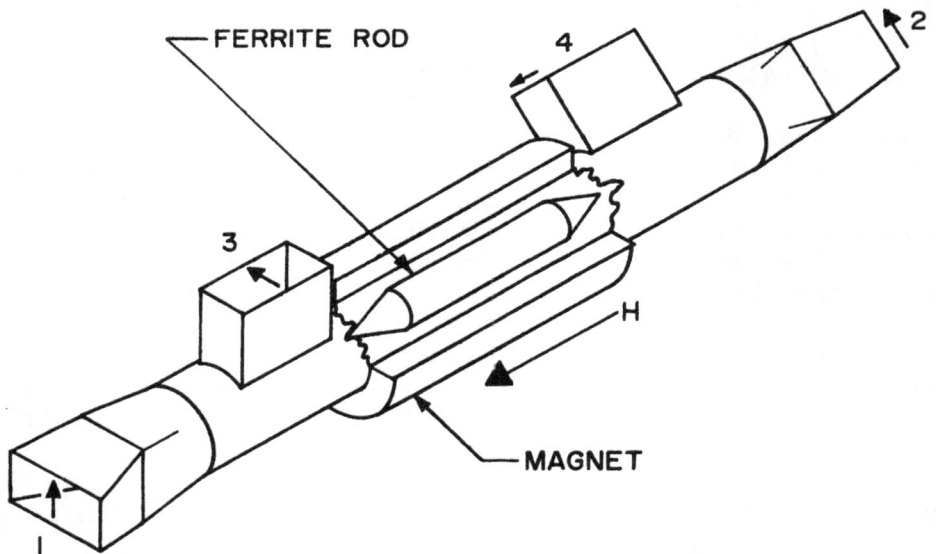

Fig. 1.8 The Faraday rotation circulator.

guides at ports 1 and 2. The circular waveguide operates in the H_{11} mode and the rectangular guides operate in the H_{10} mode.

A signal entering the device at port 1 is rotated by 45 degrees in the ferrite rod and appears at port 2. Because of the locations of the waveguides at ports 3 and 4, the mode of operation of these guides, and the direction of the electric vector of the signal entering at port 1, little or no signal is coupled to ports 3 and 4. For the same reasons, signals incident on port 3 are coupled only to port 4 and signals entering port 4 appear only at 1. Thus, the circulator is a four-port device with the signal path 1–2–3–4–1–. . .

Although Faraday rotation circulators are seldom (if ever) used in modern systems, the theory of operation interestingly follows from that of the optical Faraday effect.

In 1949, Polder [8] opened the door to extensive research involving microwave ferrites when he mathematically described the permeability of ferrites as a function of magnetization, microwave frequency, and properties of the ferrite material.

A tensor is an abstract object representing a generalization of the vector concept. If we have a product of two vectors, **AB**, also called a dyad, we can form a new vector, **D**, from the dyad **AB** and another vector **C** by applying the dot product:

$$\mathbf{D} = (\mathbf{AB}) \cdot \mathbf{C} = \mathbf{A}(\mathbf{B} \cdot \mathbf{C})$$

The dyad has properties that are different from those of either a vector or a scalar. It can be described as an operator, a linear vector operator in particular. **D** is a linear function of vector **C**.

A linear vector operator is also called a tensor. The name *tensor* was first applied when linear vector functions were used in writing elastic deformations and the corresponding stress and strain relations.

Although tensors may be composed of any number of vectorial factors and dimensions, the most commonly occurring tensors are of rank 2 (the product of two vectors) and are in three-dimensional space. These tensors have nine components, which we can write in much the same way we write vector components:

$$
\begin{aligned}
\boldsymbol{T} = {} & T_{xx}\hat{x}\hat{x} + T_{xy}\hat{x}\hat{y} + T_{xz}\hat{x}\hat{z} \\
& + T_{yx}\hat{y}\hat{x} + T_{yy}\hat{y}\hat{y} + T_{yz}\hat{y}\hat{z} \\
& + T_{zx}\hat{z}\hat{x} + T_{zy}\hat{z}\hat{y} + T_{zz}\hat{z}\hat{z}
\end{aligned}
$$

where $\boldsymbol{T}$ represents the general tensor. We can also write the components in matrix form:

$$T = \begin{pmatrix} T_{xx} & T_{xy} & T_{xz} \\ T_{yx} & T_{yy} & T_{yz} \\ T_{zx} & T_{zy} & T_{zz} \end{pmatrix}$$

The permeability of magnetized ferrite material is described by a tensor.

Tensors have special characteristics that enable them to undergo certain types of transformations under changes of coordinate system. There are also operations that can be performed on tensors such as transposition and diagonalization. A complete treatment of tensor algebra is beyond the scope of this book.

The physical significance of the tensor permeability is that the **B** and **H** vectors may point in different directions.

We begin our derivation of the Polder tensor with the Gilbert equation of motion of the magnetization, which bears some resemblance to Equation (1.4) except that it includes a damping factor, α:

$$\frac{d\mathbf{M}}{dt} = \gamma(\mathbf{M} \times \mathbf{H}) + \frac{\alpha \mathbf{M}}{\mathbf{M}} \times \frac{d\mathbf{M}}{dt} \tag{1.8}$$

The damping factor is given by

$$\alpha = \frac{\Delta H \gamma}{2\omega_0} \tag{1.9}$$

where ΔH is the resonance line width of the ferrite material. H and M in Equation (1.8) are defined as

$$\mathbf{H} = \hat{z}H_{dc} + \mathbf{h}e^{j\omega t} \tag{1.10}$$

$$\mathbf{M} = \hat{z}M_0 + \mathbf{m}e^{j\omega t} \tag{1.11}$$

Here, M_0 is the ferrite magnetization and **h** and **m** are the microwave magnetic field and magnetization components, respectively. If we substitute equations (1.10) and (1.11) into Equation (1.8), we derive

$$j\omega\mathbf{m} = \gamma M_0(\hat{z} \times \mathbf{h}) + (\omega_0 + j\omega\alpha)(\hat{z} \times \mathbf{m}) \tag{1.12}$$

We now decompose Equation (1.12) into x, y, and z components:

$$j\omega m_x = -(\omega_0 + j\omega\alpha)m_y - \gamma\mathbf{M}_0 h_y \tag{1.13}$$

$$j\omega m_y = \gamma\mathbf{M}_0 h_x + (\omega_0 + j\omega\alpha)m_x \tag{1.14}$$

$$jwm_z = 0 \tag{1.15}$$

We solve for m_x and m_y and write

$$m_x = - \frac{(\omega_0 + j\omega\alpha)\gamma M_0}{(\omega_0 + j\omega\alpha)^2 - \omega^2} h_x - \frac{j\omega\gamma M_0}{(\omega_0 + j\omega\alpha)^2 - \omega^2} h_y \tag{1.16}$$

$$m_y = \frac{j\omega\gamma M_0}{(\omega_0 + j\omega\alpha)^2 - \omega^2} h_x - \frac{(\omega_0 + j\omega\alpha)\gamma M_0}{(\omega_0 + j\omega\alpha)^2 - \omega^2} h_y \tag{1.17}$$

To simplify these expressions, we use the substitutions:

$$\chi = - \frac{(\omega_0 + j\omega\alpha)\gamma M_0}{(\omega_0 + j\omega\alpha)^2 - \omega^2} \tag{1.18}$$

$$\kappa = \frac{\omega\gamma M_0}{(\omega_0 + j\omega\alpha)^2 - \omega^2} \tag{1.19}$$

The simplification of Equations (1.15), (1.16), and (1.17) gives us

$$m_x = \chi h_x - j\kappa h_y \tag{1.20}$$

$$m_y = j\kappa h_x + \chi h_y \tag{1.21}$$

$$m_z = 0 \tag{1.22}$$

These equations can be rewritten in matrix form:

$$\begin{pmatrix} m_x \\ m_y \\ m_z \end{pmatrix} = \begin{pmatrix} \chi & -j\kappa & 0 \\ j\kappa & \chi & 0 \\ 0 & 0 & 0 \end{pmatrix} \begin{pmatrix} h_x \\ h_y \\ h_z \end{pmatrix} \tag{1.23}$$

An intermediate result in our derivation of the Polder tensor is the susceptibility tensor, which is actually a factor in Equation (1.23):

$$\underline{\chi} = \begin{pmatrix} \chi & -j\kappa & 0 \\ j\kappa & \chi & 0 \\ 0 & 0 & 0 \end{pmatrix} \tag{1.24}$$

χ and κ are complex variables, $\chi = \chi' + j\chi''$ and $\kappa = \kappa' + j\kappa''$. The real and imaginary components, derived from Equations (1.18) and (1.19), are given by

$$\chi' = -\frac{\gamma M_0 \omega_0 [\omega_0^2 - \omega^2(1 - \alpha^2)]}{[\omega_0^2 - \omega^2(1 + \alpha^2)]^2 + 4\omega^2\omega_0^2\alpha^2} \qquad (1.25)$$

$$\chi'' = -\frac{\gamma M_0 \omega\alpha [\omega_0^2 + \omega^2(1 + \alpha^2)]}{[\omega_0^2 - \omega^2(1 + \alpha^2)]^2 + 4\omega^2\omega_0^2\alpha^2} \qquad (1.26)$$

$$\kappa' = \frac{\gamma M_0 \omega [\omega_0^2 - \omega^2(1 + \alpha^2)]}{[\omega_0^2 - \omega^2(1 + \alpha^2)]^2 + 4\omega^2\omega_0^2\alpha^2} \qquad (1.27)$$

$$\kappa'' = \frac{2\gamma M_0 \omega_0 \omega^2 \alpha}{[\omega_0^2 - \omega^2(1 + \alpha^2)]^2 + 4\omega^2\omega_0^2\alpha^2} \qquad (1.28)$$

Because the permeability $\mu = 1 + \chi$, we can make slight modifications to the susceptibility tensor (1.24) to arrive at the Polder permeability tensor:

$$\underline{\mu} = \begin{pmatrix} \mu & -j\kappa & 0 \\ j\kappa & \mu & 0 \\ 0 & 0 & 1 \end{pmatrix} \qquad (1.29)$$

In most cases, α^2 is much less than unity, so the $\omega^2(1 - \alpha^2)$ terms in Equations (1.25) to (1.28) can be changed to simply ω^2. The elements of the permeability tensor now are

$$\mu' = 1 - \frac{\gamma M_0 \omega_0 (\omega_0^2 - \omega^2)}{(\omega_0^2 - \omega^2)^2 + 4\omega^2\omega_0^2\alpha^2} \qquad (1.30)$$

$$\mu'' = -\frac{\gamma M_0 \omega\alpha(\omega_0^2 + \omega^2)}{(\omega_0^2 - \omega^2)^2 + 4\omega^2\omega_0^2\alpha^2} \qquad (1.31)$$

$$\kappa' = \frac{\gamma M_0 \omega(\omega_0^2 - \omega^2)}{(\omega_0^2 - \omega^2)^2 + 4\omega^2\omega_0^2\alpha^2} \qquad (1.32)$$

$$\kappa'' = \frac{2\gamma M_0 \omega_0 \omega^2 \alpha}{(\omega_0^2 - \omega^2)^2 + 4\omega^2\omega_0^2\alpha^2} \qquad (1.33)$$

Now that we have equations for the permeability of the ferrite material, we can move on to find propagation constants for microwaves incident on ferrites. We assume that a wave makes an angle θ with the z axis, which is also the axis of the applied dc magnetic field, and an angle ϕ with the x axis, as shown in Figure 1.9.

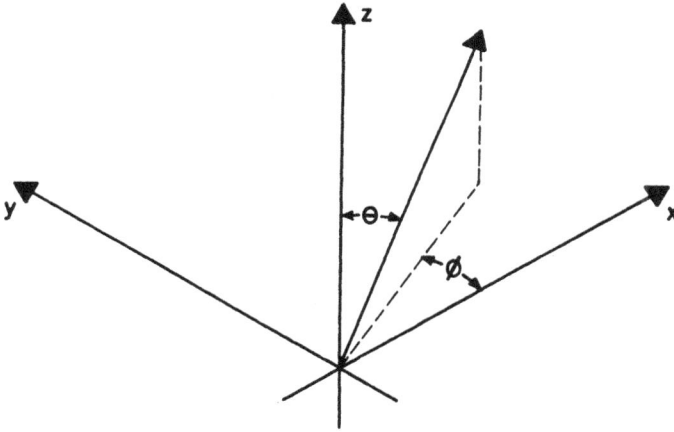

Fig. 1.9 Direction of wave propagation in ferrite.

From Maxwell's equations, it can be shown that

$$\mathbf{b} = \frac{\Gamma^2}{\varepsilon(j\omega)^2} [h_x - (\sin\theta\cos\phi\, h_x + \sin\theta\sin\phi h_y \tag{1.34}$$
$$+ \cos\theta\, h_z)\sin\theta\cos\phi]$$

where Γ is the propagation constant and ε is the permittivity. We can also equate b, the magnetic induction, to the permeability tensor:

$$\mathbf{b} = \underline{\mu}_0 \begin{pmatrix} \mu & -j\kappa & 0 \\ j\kappa & \mu & 0 \\ 0 & 0 & 1 \end{pmatrix} \begin{pmatrix} h_x \\ h_y \\ h_z \end{pmatrix} \tag{1.35}$$

Equating the right sides of Equations (1.34) and (1.35), we solve for Γ, given by

$$\Gamma_\pm = j\omega(\mu_0\varepsilon)^{1/2}$$
$$\left(\frac{(\mu^2 - \mu - \kappa^2)\sin^2\theta + 2\mu \mp [(\mu^2 - \mu - \kappa^2)^2\sin^4\theta + 4\kappa^2\cos^2\theta]^{1/2}}{2[(\mu - 1)\sin^2\theta + 1]} \right)^{1/2} \tag{1.36}$$

where Γ_+ and Γ_- are the propagation constants for waves that travel in the z direction. The plus and minus indicate different polarizations of the elliptically polarized waves. These waves propagate with different velocities through the ferrite. The propagation constants do not depend on the angle the waves make with the x axis.

The propagation constant describes the propagation of an electromagnetic wave in a particular medium. For a homogeneous, isotropic medium (magnetized ferrites are not isotropic), the propagation constant is given by

$$\Gamma = \pm j\omega(\mu_0\mu\varepsilon)^{1/2}\left(1 - \frac{j\sigma}{\omega\varepsilon}\right)^{1/2}$$

where ε is the absolute permittivity for the medium and σ is the conductivity.

Γ is a complex quantity that can be broken down into an attenuation term (α) and a phase term (β):

$$\Gamma = \alpha + j\beta$$

By binomial theorem expansion, we can separate the attenuation and phase terms. The first few terms of the expansion give us:

$$\alpha = \frac{\sigma}{2}\left(\frac{\mu_0\mu}{\varepsilon}\right)^{1/2}$$

$$j\beta = j\omega(\mu_0\mu\varepsilon)^{1/2}\left[1 + \frac{1}{8}\left(\frac{\sigma}{\omega\varepsilon}\right)^{1/2}\right]$$

The attenuation term (α) is a measure of the change in amplitude of the electric or magnetic field intensity per unit length in the path of propagation. The phase term (β) is a measure of the change in phase of the electric or magnetic field vectors per unit length.

If the conductivity of the medium is very low, the phase term reduces to

$$j\beta = j\omega(\mu_0\mu\varepsilon)^{1/2}$$

and the attenuation term becomes negligible. Thus, we can approximate the propagation constant with the phase term or vice versa if the attenuation term is small.

The units for the attenuation and phase terms are nepers per unit length and radians per unit length, respectively.

To show the usefulness of the propagation constants in more detail, we will consider some special cases: the transverse-field case, where the dc magnetic field is perpendicular to the direction of propagation; the longitudinal field case, where the dc magnetic field is parallel to the direction of propagation, not the case of Faraday rotation; and the case of resonance absorption.

When the direction of propagation is transverse to the applied dc magnetic field, $\theta = \pi/2$ and Equation (1.36) reduces to

$$\Gamma_+ = j\omega(\mu_0\varepsilon)^{1/2} \left(\frac{\mu^2 - \kappa^2}{\mu}\right)^{1/2} \tag{1.37}$$

$$\Gamma_- = j\omega(\mu_0\varepsilon)^{1/2} \tag{1.38}$$

It is interesting to note that in this transverse-field case h_x and h_y have the relation:

$$h_y = -j\frac{\kappa}{\mu} h_x \tag{1.39}$$

and that b is linearly polarized transverse to the z axis.

The transverse case applies to junction circulators, and the factor κ/μ in Equation (1.39) has been used [9] in various theoretical studies to indicate specific ferrite magnetic operating points.

In a junction circulator, the dc magnetic field is applied to the ferrite so that the direction of RF propagation is transverse to the dc magnetic field. Two elliptically polarized waves with propagation constants given by Equations (1.37) and (1.38) are present in the ferrite. If the ferrites are in the form of disks, the two waves rotate in the disks in opposite directions. They are periodically in and out of phase depending on the position around the disk perimeter (assuming the disks have the dc magnetic field applied parallel to their axes). If we provide a means of coupling RF energy into and out of these nodes and antinodes, we have a junction circulator. The operation of the junction circulator is described in more detail in Section 5.1.

In the longitudinal field case $\theta = 0$ and Equation (1.36) reduces to

$$\Gamma_\pm = j\omega[\mu_0(\mu \mp \kappa)\varepsilon]^{1/2} \tag{1.40}$$

Certain assumptions can be made which simplify the expressions for $\mu - \kappa$ and $\mu + \kappa$. In the magnetic region where the ferrite is unsaturated, that is, $\omega \gg \omega_0$ and $\alpha^2 \ll 1$, appropriate simplifications can be applied to Equations (1.30) to (1.33). Terms involving α^2 in Equations (1.25) to (1.28) can be dropped except when the ferrite is being used in a resonance device or the microwave frequency is very close to the ferrimagnetic resonance frequency. For devices that operate far from resonance in the frequency domain, it is possible that $\alpha \ll 1$. If this is the case, terms involving α may be dropped.

The amount of rotation caused by a longitudinally magnetized ferrite in a Faraday rotation device is given by

$$\phi = \left(\frac{\beta_+ - \beta_-}{2}\right) L \tag{1.41}$$

where L is the length of the ferrite. Equation (1.41) contains the average of the rotations caused by two elliptically polarized waves. These two waves have phase constants given by β_+ and β_-. For the typical case of a low value of applied dc magnetic field, β_+ and β_- can be approximated by Γ_+ and Γ_-. This is due to the low value of the attenuation-constant component of Γ.

In order to capitalize on the ferrimagnetic resonance absorption effect, we must again consider two counter-rotating elliptically polarized waves. The wave that rotates in the same direction as the precession of the electron spin moments will suffer high attenuation if the microwave frequency is at or near the ferrimagnetic resonance frequency. The other rotating wave will not suffer this attenuation.

Our treatment of the theory of microwave propagation in ferrites will not be complete until we discuss spin waves. If a microwave magnetic field is applied to a ferrite that is also biased by a dc magnetic field, the magnetization vector of the ferrite will be aligned with the dc magnetic field at very low RF signal levels. As the RF signal level is increased, the magnetization vector may begin to move away from alignment with the dc field. It is impossible, however, for the magnetization in the direction of the dc field to drop to zero or reverse polarity because of spin-wave effects.

As the critical microwave magnetic field, H_c, is reached, the change in angle of the magnetization vector excites spin waves. There is a spatial periodicity to the amount the magnetization vector deviates from the equilibrium direction due to nonuniformity of the magnetic fields, thermal excitation of the ferrite atoms (at 0 K it might be possible to eliminate spin-wave effects), and boundary effects caused by the dimensions of the ferrite sample. This spatial periodicity is the period of the spin wave. The name *spin wave* is derived from the electron spin axis direction, which is in the direction of the magnetization vector at equilibrium, and the sinusoidal wave nature that the spin waves possess.

Spin waves, mutually coupled and propagating in all directions in the ferrite, cause a broadening and reduction in amplitude of the ferrimagnetic resonance. Spin waves having half the frequency of the waves that cause the broadening of the main ferrimagnetic resonance are responsible for a subsidiary resonance.

The spin waves are most troublesome in the design of high-power, below-resonance devices. These devices usually operate in the region between ferrimagnetic resonance and the subsidiary resonance, so that a broadening of either resonance leads to increased insertion loss at high power.

The theoretical information presented in this chapter is mainly for qualitative purposes at this point. The mathematical complexity of the equations, together with the fact that none of the equations will yield high-performance circulator designs without the addition of other factors, prevents their quantitative use.

The information does, however, form a theoretical base for further development in later chapters, particularly Chapter 5.

REFERENCES

1. Waldron R. A., *Ferrites: An Introduction for Microwave Engineers* (London: Van Nostrand, 1962).
2. Keffer, F., "The Magnetic Properties of Materials," *Materials* (San Francisco: W. H. Freeman, 1967).
3. Wert, C. A., and R. M. Thomson, *Physics of Solids* (New York: McGraw-Hill, 1970).
4. Soohoo, R. F., *Microwave Magnetics* (New York: Harper and Row, 1985).
5. Symon, K. R., Mechanics (Reading, MA: Addison-Wesley, 1971).
6. Riches, E. E., *Ferrites: A Review of Materials and Applications* (London: Mills and Boon, 1972).
7. Jenkins, F. A., and H. E. White, *Fundamentals of Optics* (New York: McGraw-Hill, 1976).
8. Polder, D., "On the Theory of Ferromagnetic Resonance," *Phil. Mag.,* Vol. 40, 1949, pp. 99–115.
9. Bosma, H., "On Stripline Y-Circulation at UHF," *IEEE Transactions on Microwave Theory and Techniques,* January 1964, pp. 61–73.

Chapter 2
Circulator Specification

2.1 THE PARAMETERS

Engineers assigned the task of designing a ferrite circulator or isolator need to know how to specify these devices. If specifications are supplied to the designer, he or she must know whether they are reasonable—whether the device can be built. The design engineer also must know how to generate specifications for circulators in order to market his or her designs.

Although this book is aimed primarily at design engineers, others may learn how to specify isolators and circulators by reading this chapter.

Circulator specification is complicated by the large variety of devices on the market. Several different types of circulators—junction, lumped-constant, differential phase shift, field displacement, resonance, and others, available in coaxial and waveguide versions that cover many frequency bands—crowd the market.

Incorrect circulator specification can lead to inconvenience at best and disaster at worst. If a particular circulator design will not perform the desired function, it is necessary to try a different design or eliminate the circulator entirely. Most circulators [1] are used to protect high-power RF sources. In this application, a circulator failure could easily destroy the power source, possibly costing thousands of dollars and lost time. Circulator failures in national defense systems could have much more serious consequences.

The first commercial microwave circulators appeared in the early 1950s [2]. Faraday rotation circulators (see Figure 1.8) were among the first to appear. These were later replaced by resonance isolators and differential phase shift circulators, which have higher power-handling characteristics and are simpler in construction. Field-displacement devices soon joined the ranks, as did junction circulators in the early 1960s. In 1964, Yoshihiro Konishi [3] brought the lumped-element circulator to the attention of the microwave industry.

A circulator is defined as a device with ports (coaxial connectors or wave-guide flanges) arranged such that energy entering a port is coupled to an adjacent port, but not coupled to the other ports. This circulator definition is depicted in Figure 2.1. A circulator can have any number of ports, but only circulators with three or more ports involve the nonreciprocal behavior of ferrites. A one-port circulator would be analogous to a short circuit. A two-port circulator would perform the same function as a section of transmission line.

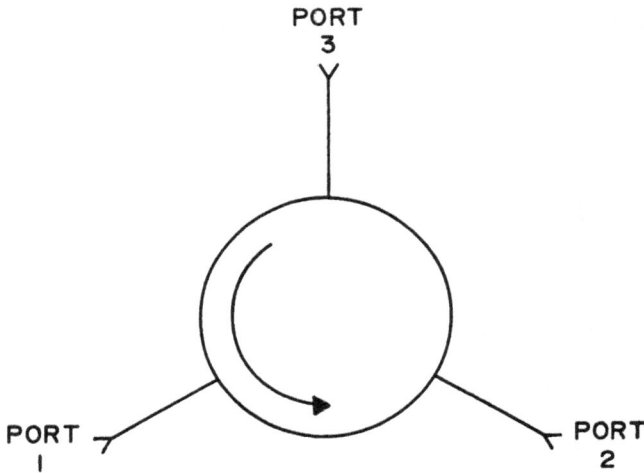

Fig. 2.1 The circulator.

It is interesting to note that if there are reflections from the port to which a signal is coupled, these reflections are in turn coupled to the next adjacent port. Thus, a signal entering port 1 in Figure 2.1 that encounters a short circuit at port 2 will be coupled to port 3. If there is also a short circuit at port 3, the signal, having gone full circle, reappears at port 1.

An isolator is a two-port device that transfers energy from input to output with little attenuation and from output to input with high attenuation. The isolator, shown in Figure 2.2, can be derived from a three-port circulator by simply placing a matched load (reflectionless termination) on one port. Junction isolators are made in this manner. Field displacement isolators and resonance isolators are strictly two-port devices and do not have third ports.

The theory of operation of circulators has been described using a water-pipe analogy [4], but because we have already discussed theory of operation in Chapter 1, we will dispense with the water piping. The theory of operation of all circulators and isolators is as described in Chapter 1. The specific applications of the theory

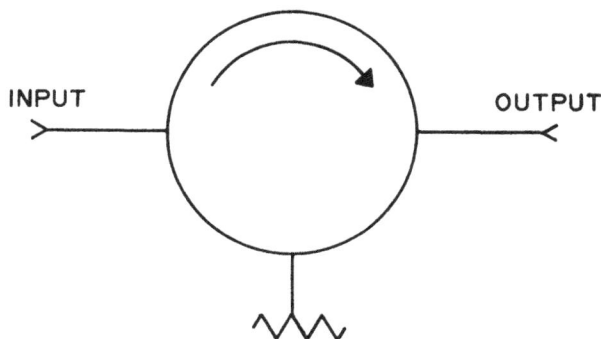

Fig. 2.2 The isolator.

are described in the sections on the particular types of circulators and isolators that appear later in this chapter.

The important circulator (and isolator) parameters are:

1. Frequency range or center frequency and bandwidth;
2. Isolation;
3. Insertion loss;
4. Voltage standing wave ratio (VSWR);
5. Power handling;
6. Linearity;
7. Temperature range (operating and storage);
8. Size and weight;
9. Shielding (electromagnetic and magnetic);
10. Other environmental factors: shock, vibration, humidity;
11. Number of ports and configuration;
12. Type of transmission medium (coaxial line, waveguide, microstrip).

In addition, isolators will have a power rating for the internal termination or a maximum VSWR specification for the load that is to be connected to the device output.

The frequency range of a circulator, usually designated by two frequencies, a lower limit and an upper limit, is defined as the range of frequencies over which all the electrical specifications apply. This frequency range is not the 3 dB bandwidth often used for other microwave components. The center frequency, f_0, is related to the lower and upper frequency limits, f_1 and f_2 by

$$f_0 = \frac{f_1 + f_2}{2} \tag{2.1}$$

The bandwidth is given by

$$BW = f_2 - f_1 \tag{2.2}$$

and the percentage bandwidth is defined as

$$\% \ BW = \frac{BW(100\%)}{f_0} \tag{2.3}$$

Isolation is the ratio of a signal power applied to the output to that measured at the input, after passing through the isolator in the reverse direction. This ratio is typically expressed in decibels. The isolation of a circulator is measured with the third port (and other ports if there are more than three) terminated in a matched load. Isolation is a function of the VSWR at the third port and the VSWR of the test termination. The VSWR of the test termination, or matched load as it is also called, should be low enough so that it will not cause a significant measurement error. If the reflection coefficient of the test termination is an order of magnitude lower than the value of reflection coefficient that would result in the specified isolation figure, the termination will not cause a significant error in measurement.

The VSWR and reflection coefficient are related by

$$VSWR = \frac{1 + \rho}{1 - \rho} \tag{2.4}$$

The isolation of a circulator and the *return loss* (RL) at the third port have a direct correspondence. That is, if the third port has a return loss of 20 dB and it is terminated in a reflectionless termination, the isolation will be measured as 20 dB (assuming insertion loss is negligible). Return loss has a simple relationship to the reflection coefficient:

$$RL = -20 \ \log_{10}(\rho) \tag{2.5}$$

Figure 2.3 shows a typical test setup for measuring isolation, VSWR, and insertion loss of circulators and isolators. A wide range of test equipment is used by microwave component manufacturers, from antique signal generators and slotted lines to automatic vector network analyzers. We will make no comment as to which test installation is the best, only that Figure 2.3 shows a middle-of-the-road setup and the equipment details are left to the test technician.

The determination of the required amount of isolation depends on the particular application of the device. Here, we will discuss the isolation needed for a circulator used to protect an RF power source from a potential high-VSWR condition.

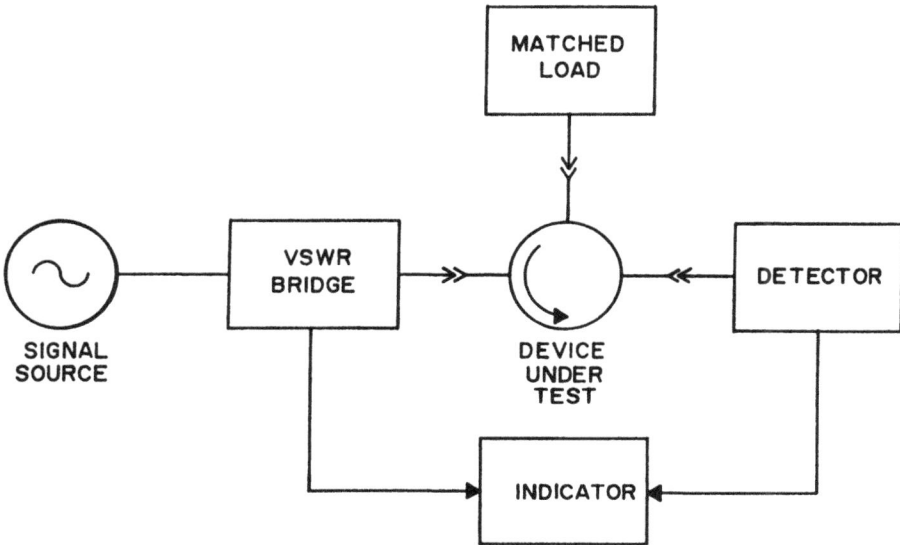

Fig. 2.3 Typical circulator test setup.

The pertinent data for this decision are: maximum VSWR to be presented by load to the circulator output; the circulator VSWR; and the maximum VSWR that can safely be presented to the RF source. We start by converting the VSWR figures to voltage reflection coefficients, because we can easily add two signals together when we work with these coefficients. We use the conversion:

$$\rho = \frac{\text{VSWR} - 1}{\text{VSWR} + 1} \tag{2.6}$$

We now represent the load reflection coefficient by ρ_L, the circulator reflection coefficient by ρ_C, and the source coefficient by ρ_S. First, the reflected signal from the load adds to the signal reflected at the circulator output port. The worst case is when the two signals are in-phase and the voltage reflection coefficients add directly. Next, this new signal traveling through the circulator toward the source is attenuated by the isolation of the circulator, denoted by I. Finally, the signal interacts with the signal reflected from the circulator input port. We again use the worst-case in-phase scenario. This final signal must be below ρ_S. The various signals in this analysis are shown in Figure 2.4.

The equation for this analysis is

$$\rho_S > (\rho_L + \rho_C)10^{(I/-20)} + \rho_C \tag{2.7}$$

Solving for I, the required isolation, we derive

$$I = -20 \log_{10}\left(\frac{\rho_S - \rho_C}{\rho_L + \rho_C}\right) \tag{2.8}$$

Figure 2.5 is a plot of the required isolation for various values of circulator VSWR versus the maximum allowable VSWR presented to the source.

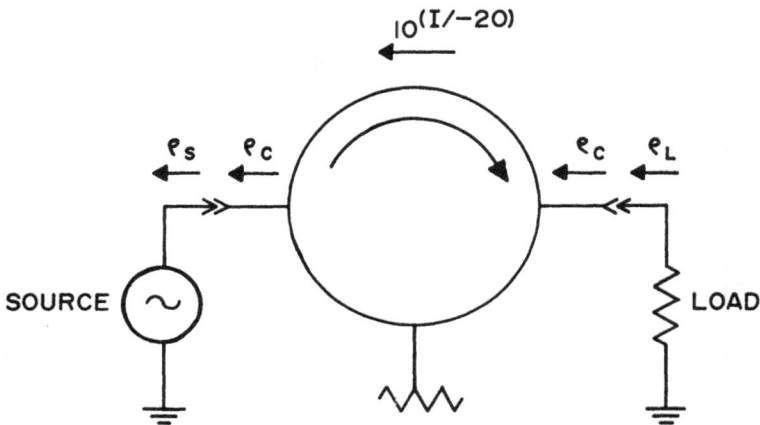

Fig. 2.4 Circulator analysis to determine required isolation.

Insofar as the voltage reflection coefficient can never exceed unity, if the load reflection coefficient is 1, the denominator in Equation (2.8) becomes 1.

If more isolation is required than the amount that can be furnished by one circulator, circulators can be cascaded to achieve the desired performance. This technique is often accomplished within the same mechanical package; however, it also increases insertion loss. Insertion loss has been neglected in Equation (2.8), as it can be in most practical situations where we are considering the effect of reflections.

Insertion loss is the ratio of output signal to input signal with the signal applied to the input, expressed in decibels. It is measured in the same manner as isolation and the third port is terminated with a matched load for this test as well. Some specifications [5] call for insertion loss to be measured in all the forward transmission paths (ports 1-2, 2-3, and 3-1 in Fig. 2.1), but this is not always necessary because we are usually only concerned with having low insertion loss in one path. Higher insertion loss in the other paths is often desirable.

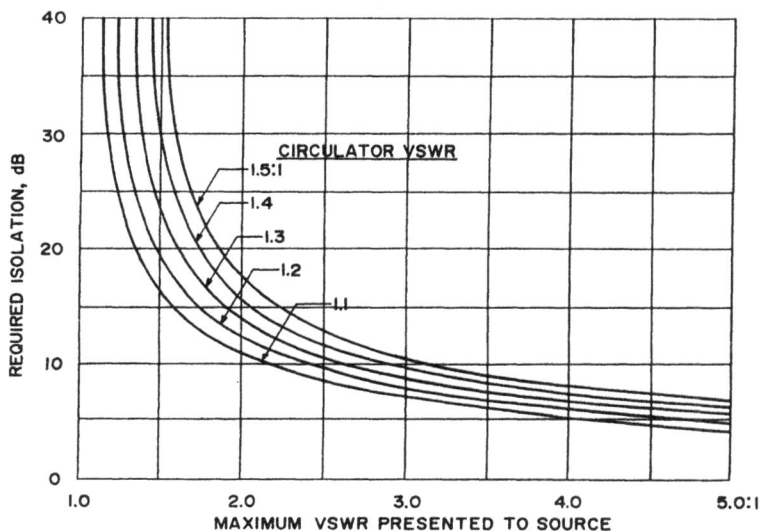

Fig. 2.5 Required isolation to achieve a given source VSWR with an infinite VSWR presented to circulator output.

It might seem obvious that circulators designed to handle high power levels will have low insertion loss to minimize the amount of power dissipated in the circulator in the form of heat, but in many cases the insertion loss of high-power circulators is higher than the loss of low-power units. This paradox will be explained further when we discuss power handling.

Factors such as conductor losses of the transmission medium, dielectric losses, and magnetic losses in the ferrite set lower limits on the insertion loss specifications of practical circulators. The expected insertion losses of several different types of circulators are given later in this chapter.

VSWR can also be measured using the setup shown in Figure 2.3. The VSWRs at all ports are usually taken. For a symmetrical circulator, the return losses at the input and output equal the isolation due to one terminated port. If several circulator junctions are cascaded to achieve higher isolation, the isolation will certainly be higher than the return losses.

Specifications for VSWR depend on the device bandwidth and design. Approximate figures are given in the sections on each type of circulator.

Circulator power handling depends on several parameters:

1. Power handling of coaxial connectors, waveguide, or another medium;
2. Power handling of the basic circulator;
3. Power threshold of ferrites.

Power handling limits can be exceeded in two modes: peak power and average power. Excessive peak power leads to corona and arcing due to the high voltages that can be present. Average power failure is typically due to overheating.

Table 2.1 lists continuous-wave (CW) power handling [6] for rectangular waveguides commonly encountered in circulator design. It is unlikely that the CW rating of any of the waveguides would be exceeded by a CW signal, but the CW ratings are useful for determining the ratings under pulsed conditions. For pulsewidths much greater than 1 μs or pulse repetition rates greater than 2 kHz, the CW power handling figures can be used for the peak power handling. With narrow pulses or low repetition rates, the peak power rating will be higher. Pulse widths on the order of 3 ns, for example, can increase the peak power handling by a factor of 8–9.

Table 2.1 Rectangular Waveguide Power Handling [6]

Waveguide Type	Frequency Range	CW Power Handling
WR 42	18.0–26.5 GHz	160 kW
WR 51	15.0–22.0	300
WR 62	12.4–18.0	440
WR 75	10.0–15.0	600
WR 90	8.2–12.4	730
WR 112	7.05–10.0	1.2 MW
WR 137	5.85–8.2	1.9
WR 159	4.9–7.05	2.7
WR 187	3.95–5.85	3.2
WR 229	3.3–4.9	5.3
WR 284	2.6–3.95	7.3
WR 340	2.2–3.3	12
WR 430	1.7–2.6	18
WR 510	1.45–2.2	25
WR 650	1.12–1.7	40
WR 770	960–1450 MHz	58
WR 1150	640–960	130
WR 1800	410–625	310
WR 2300	320–490	510

The mechanism for peak power breakdown in a waveguide is the production of free electrons and collisions of the accelerated free electrons with gas molecules, leading to the production of more free electrons. Eventually, enough electrons are present to cause arcing or breakdown. It takes time for the quantity of free electrons to build up, which is why short pulses and low repetition rates are not as troublesome as long pulses and high repetition rates.

The breakdown voltages of the waveguides can be increased if the mean free paths of the electrons can be shortened. This can be done by pressurizing the waveguide. Also, substitution of other dielectric gases for air can be helpful in preventing breakdown.

The maximum VSWR to be present at the connections to the circulator should also be considered. The voltage is increased by a factor of $1 + \rho$:

$$V \propto 1 + \rho \tag{2.9}$$

The peak power ratings of coaxial connectors can be determined [7] from the 60 Hz dielectric withstanding voltage using

$$P = \frac{V_{dw}^2}{Z_0(1 + \rho)^2 F^{1/2} A} \tag{2.10}$$

where V_{dw} is the connector dielectric withstanding voltage, Z_0 is the characteristic impedance, F is the frequency in GHz, and A is an altitude derating factor. Tables 2.2 and 2.3 list dielectric withstanding voltages for some common connectors and A values, respectively.

The average or CW power handling of coaxial connectors is usually equal to the average power capability of the associated cable. Figure 2.6 shows the average power handling versus frequency for some selected flexible cables and rigid EIA-type coaxial transmission lines.

The power handling of a circulator cannot exceed the capacity of the waveguides or connectors that interface with it, but in many cases the basic circulator will limit the power handling. Waveguides have higher power handling than coaxial transmission lines, but at low frequencies they become very large. The choice between waveguide and coaxial transmission line will normally be dictated to the circulator designer, but in some instances it may be advantageous to make transitions between the two media to improve power handling, electrical performance, or to make the transition part of the circulator (e.g., waveguide input and coaxial output).

An introduction to the three possible magnetic operating regions for circulators and isolators will make the following discussion of power threshold more clear. The magnetic operating regions, shown later in Figure 4.2, refer to the

Table 2.2 Connector Dielectric Withstanding Voltage [8]

Connector Type	Dielectric Withstanding Voltage (kV)	Mating
SMA	1	1/4-36 Thread
BNC	1.5	Bayonet
TNC	1.5	7/16-28 Thread
N	2.5	5/8-24 Thread
SC	3	11/16-24 Thread
HN	5	3/4-20 Thread
LC	10	1 1/4-18 Thread

Table 2.3 Altitude Derating Factors [7]

Altitude (ft)	Derating Factor (A)
0	1
10,000	2
20,000	5
30,000	7
40,000	10
50,000	12
60,000	15
70,000	20
80,000	22

magnitude of the applied dc magnetic field relative to the amount of field required for ferrimagnetic resonance. Below resonance means simply that the amount of dc magnetic field is below that required for resonance, and above resonance means the field is above the resonance field magnitude. Thus, the three magnetic operating regions are below resonance, resonance, and above resonance. The resonance

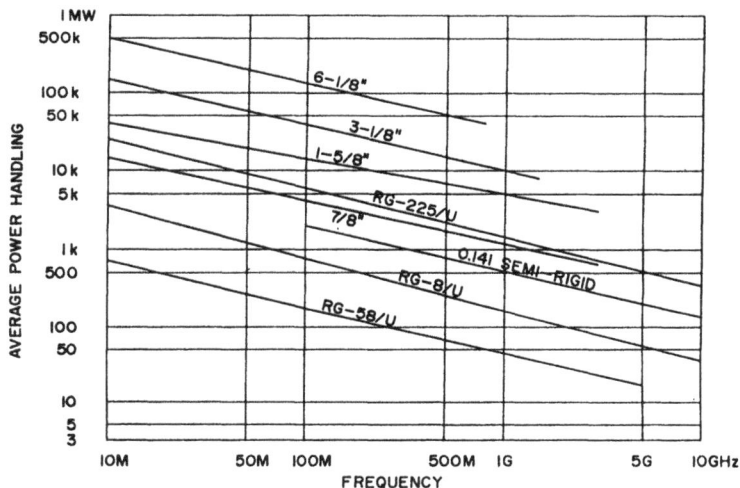

Fig. 2.6 Average power handling of selected coaxial transmission lines.

region is used only for resonant devices such as resonance isolators. Circulators, in general, fall into either the below- or above-resonance category.

The power threshold of the ferrites is of concern when the specifications indicate that a below-resonance design must be used. When a certain critical power level is reached, the excitation of spin waves begins as described in Chapter 1. When this happens, the insertion loss of the circulator increases, hence we have a nonlinearity. This nonlinearity causes the generation of harmonics and other mixing products if two or more signals are present.

Circulators for communications systems may have either third-order inter-modulation levels or third-order intercept points specified. A third-order intercept point of +95 dBm has been suggested [1] as attainable for circulators operating at frequencies up to about 4 GHz. Reasonably linear performance can be obtained [9] at power levels up to 30 kW at 2 GHz. For octave-bandwidth circulators, 2 kW at 2 GHz and 200 W at 16 GHz are typical figures.

Increased power threshold and linearity can be obtained by changing the ferrite material and geometry. Broad-bandwidth and high power threshold are contradictory terms, as are low loss and high power threshold. Ironically, at higher power levels where you would think low loss would be a necessity, we must use ferrite materials that have greater spin-wave line widths with higher magnetic losses. Broad bandwidth and high power threshold are contradictory because broad bandwidth indicates the use of the below-resonance operating region, where the excitation of spin waves at high power levels causes nonlinearity. The compromises involved in ferrite material selection are discussed further in Section 4.1.

The storage temperature range of circulators can readily be extended to cover −55° to +125°C. The operating temperature range may not be as broad. Below-resonance circulators for low frequencies (below 1 GHz) contain ferrite materials that have low Curie temperatures and thus have narrow operating temperature ranges, e.g., 0° to +50°C. Curie temperature is the temperature above which the ferrite material has effectively no magnetic properties. At higher frequencies and for above resonance units, the temperature range can be increased to −55° to +85°C or even +125°C, depending on the required electrical performance. Circulators with tighter electrical specifications or broader bandwidths necessarily have reduced operating temperature ranges.

The size and weight of circulators is usually dictated by the design type. However, certain things can be done to reduce the circulator weight, such as selecting lightweight materials. Size can sometimes be reduced by incorporating dielectrics with higher relative permittivities.

Circulators have inherently good electromagnetic (*radio frequency interference* (RFI) and *electromagnetic interference* (EMI)) shielding. Leakage on the order of 30 dB is typical, and 80–100 dB is possible if attention is given to the reduction of radiation.

Almost any degree of magnetic shielding can be achieved. Because circulators contain strong permanent magnets, concern stems from the effect of the internal magnetic field on other nearby components. Furthermore, the performance of the circulator may degrade if it is brought either near or in contact with ferrous materials. Magnetic shielding may be specified as a minimum distance the circulator must be kept from ferrous material or as a maximum flux density at a given distance from the circulator.

Various other environmental factors can be specified for circulators such as vibration, shock, and humidity. The specifications depend on the environment in which the circulator is intended to be used. Military Specification MIL-C-28790 states that circulators are to be tested for shock, vibration, and moisture resistance per MIL-STD-202 methods 213, 204, and 106, respectively.

We can configure circulators and isolators with any number of ports, and package them in a wide variety of orientations and combinations. To specify the number of ports, configuration, and packaging for a device, we need to consider the particular application.

The reliability of circulators and isolators is quite high—a circulator operating in a benign environment could have a *mean time between failures* (MTBF) of 10^7 hours. Table 2.4 lists MTBF figures for circulators operating in various environments, calculated from data given in MIL-HDBK-217D [10]. The figures in Table 2.4 apply for circulators used at power levels less than or equal to 100 W. For higher-power units, the MTBF values should be halved.

Table 2.4 Circulator Reliability [10]

Environment	Mean Time Between Failures (hours)
Ground, benign	10,000,000
fixed	4,170,000
mobile	1,140,000
Space flight	10,000,000
Manpack	1,300,000
Naval, sheltered	1,610,000
unsheltered	833,000
undersea	769,000
submarine	2,700,000
hydrofoil	833,000
Airborne, transport, inhabited	2,220,000
fighter, inhabited	1,000,000
transport, uninhabited	1,540,000
fighter, uninhabited	667,000
Helicopter	588,000
Missile launch	385,000
Cannon launch	22,200
Undersea launch	435,000
Missile flight	1,280,000
Missile flight, air-breathing	909,000

2.2 JUNCTION CIRCULATORS

The most common circulators are the junction type, available in coaxial, microstrip, and waveguide varieties. Some typical junction circulators are shown in Figure 2.7(a)–(c).

Coaxial (stripline) circulators are constructed as shown in Figure 2.8. The center conductor is sandwiched between two ferrite disks or triangles. These ferrites are then placed between ground planes and magnetically biased by permanent magnets outside the ground planes. Microstrip circulators are similar, but the conductor is usually a metalization applied to the single ferrite disk. The connections to the external circuitry are made with tabs. The ground plane is simply a

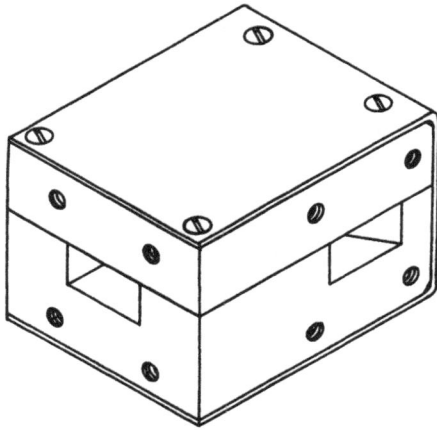

Fig. 2.7(a) A typical waveguide junction circulator.

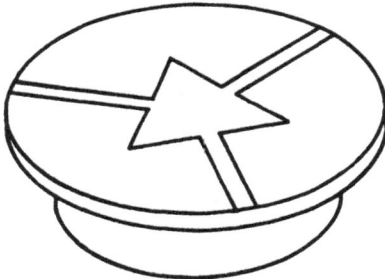

Fig. 2.7(b) A typical microstrip junction circulator.

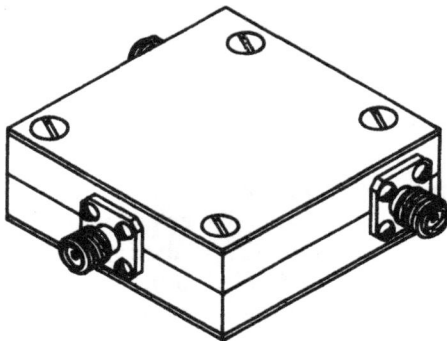

Fig. 2.7(c) A typical coaxial junction circulator.

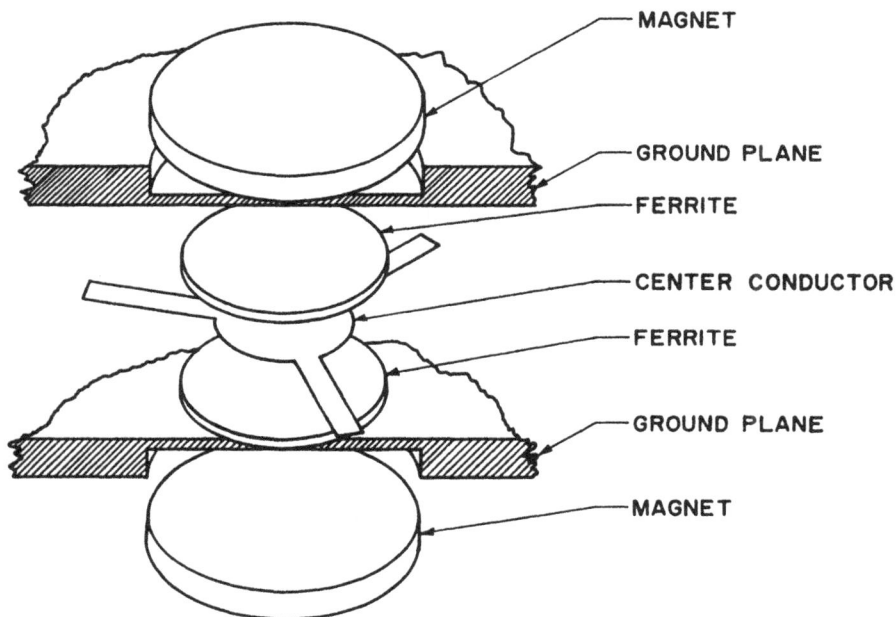

Fig. 2.8 Stripline circulator construction.

metal disk or metalization on the ferrite, and a magnet is soldered or glued to the ground plane. Screw clamps or soldering hold the circulators in place.

Waveguide junction circulators consist of sections of waveguides that form an *H*-plane junction. The ferrite disks or triangles are located against the walls of the waveguides in the *H* plane. The biasing magnets are positioned outside the waveguide.

More information about junction circulators is presented in section 5.1.

Most common rectangular waveguide bands can be covered by waveguide junction circulators. Typically, insertion loss is 0.2 dB or less and VSWR is on the order of 1.2:1. Octave bandwidths can be accommodated by double-ridged waveguide junction circulators with somewhat reduced performance.

Coaxial and microstrip circulators operate in one of two magnetic bias regions: above resonance or below resonance. The majority of waveguide units operate below resonance. Above resonance coax junctions can cover bandwidths up to about 40% with less than 0.5 dB insertion loss and 1.3:1 VSWR. Improved performance can be obtained over more narrow bandwidths, but for identical frequency ranges, waveguide junctions will have lower loss.

Below resonance, 90% bandwidths can be achieved. Octave-bandwidth units are fairly easy to construct, and may have less than 0.4 dB loss and 1.2:1

VSWR. The below-resonance designs are only used for frequencies above about 500 MHz because the required ferrite saturation magnetization becomes vanishingly small below this frequency. Devices for frequencies below 1 GHz suffer lesser performance than higher-frequency units because of the ferrite material limitations. The performance of both above- and below-resonance circulators will be discussed in more detail in Chapter 5.

Above-resonance circulators have almost no lower frequency limit but practical considerations of cost and size set a lower limit in the neighborhood of 100 MHz. At higher frequencies, above-resonance operation demands very high magnetic fields. The upper frequency limit for above-resonance circulators is between 2.5 and 3.0 GHz.

The peak power handling capability of above-resonance circulators is limited by the ground-plane spacing and connector capacity for coaxial devices and by the waveguide power capacity for waveguide devices. Increasing the ground-plane spacing of a coaxial unit will not only increase the power handling, but could change the junction impedance to a less than optimum value for broad-bandwidth performance. For this reason, coaxial circulators with very high peak power ratings must have narrow bandwidths.

The upper limit of peak power handling for below-resonance circulators is set by the onset of spin-wave propagation and the resulting increase in insertion loss.

Average power-handling capacity of above-resonance circulators is usually higher than that of below-resonance circulators because the ferrites used have higher Curie temperatures. In addition, high average powers can aggravate the nonlinear behavior of below-resonance units.

Cooling is a very important consideration for circulators that will handle high average powers. Power lost in the circulator junction due to insertion loss must be dissipated in the form of heat. If the circulator in question must pass the RF power in two paths, as it would if there were a short circuit at the circulator output, the amount of dissipation will double. The type of cooling to be employed depends on the amount of heat that must be removed from the ferrite junction. Insofar as the power density in the ferrites must be held to a maximum value, the size of the ferrites will also have some influence on the type of cooling to be used. Cooling methods commonly used, in order of increasing efficiency, are radiation, conduction, natural convection, forced air, and liquid cooling.

Junction circulators are the smallest and lightest next to lumped-constant circulators. Above-resonance devices are smaller than below-resonance units for a given frequency range. The sizing of ferrite junctions will be discussed in more detail in Chapter 5.

2.3 LUMPED-CONSTANT CIRCULATORS

The smallest circulators for the VHF and UHF bands are of the lumped-constant (also called lumped-element) design. The construction of this type of circulator is illustrated in Figure 2.9. The lumped-element circulator consists of a ferrite disk with three coils wound on it so that the RF magnetic fields are oriented at 120 degrees with respect to each other. The ferrite disk and center conductor (coil) are shown in Figure 2.9. More details regarding the construction of lumped-element circulators are presented in Chapter 5. Most lumped-constant circulators are for use with coaxial transmission lines.

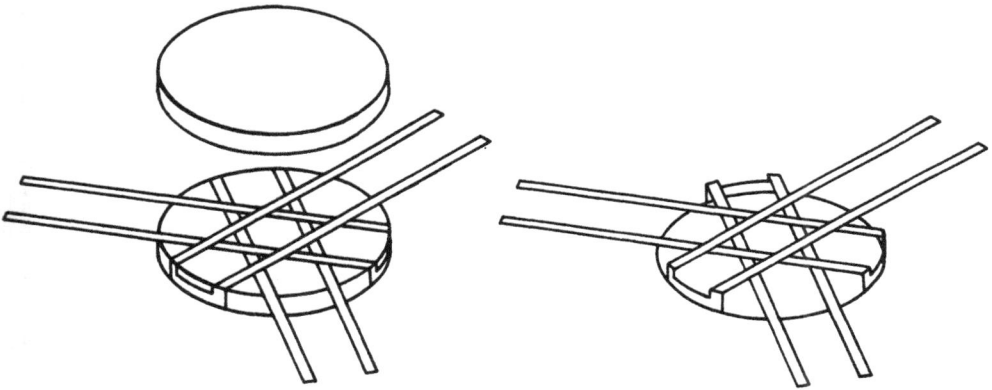

Fig. 2.9 Lumped-element circulator construction. Center conductor with ferrites (left) and without ferrites (right).

Lumped-constant circulators are most often used for frequencies below 1 GHz. Above this frequency, junction circulators are not much larger and offer lower insertion loss and simpler construction. There is practically no lower frequency limit for the operation of lumped-constant circulators; devices have been built for use at frequencies below 10 MHz.

We can accommodate bandwidths of 15% to 20% at 0.6 dB insertion loss and 1.2:1 VSWR points. With proper impedance-matching circuit design, 2 dB bandwidths of about 50% can be realized while maintaining 1.2:1 VSWR.

The power handling of lumped-constant circulators is not as high as that for junction circulators. The ferrites used are smaller and the losses are higher, so the power dissipation in the ferrites is higher. In the VHF range, 1 kW of average power can be handled with forced-air cooling. The peak power capability of lumped units is low because of the necessarily small spacing between portions of the center conductor mesh connected to adjacent ports.

Because lumped-constant circulators usually operate in the above-resonance region, there is no concern about harmonic generation or intermodulation products unless the power level is high.

In order to generate harmonics or intermodulation products, we must have a nonlinear component of some type. If there is no nonlinear component, there is no harmonic generation. Below-resonance circulators can become nonlinear if they are operated at a power level sufficient to cause spin-wave excitation. The spin waves cause increased insertion loss at high powers as described in Chapter 1. Above-resonance circulators, on the other hand, are not subject to increases in insertion loss due to spin-wave excitation. Above-resonance circulators behave as linear components.

It is interesting to note two items. First, any component can be shown to be nonlinear if a sufficiently high power level is applied to it. There are, of course, large differences in the amount of nonlinearity exhibited by "linear" and "nonlinear" components. Second, ferrite devices are manufactured that capitalize on nonlinear properties of ferrites, such as limiters and frequency multipliers.

The operating temperature range is influenced not only by the ferrite material used but also by the lumped components (capacitors and inductors) used in the matching circuits between the ferrite and the ports. Careful capacitor selection can compensate for variations in the matching circuits, and can also help reduce the effects of changes in the ferrite properties with temperature.

Lumped-constant circulators are generally not well suited for applications where they will be exposed to high levels of mechanical shock or vibration due to the instability of the lumped-element matching circuits.

2.4 DIFFERENTIAL PHASE SHIFT CIRCULATORS

The differential phase shift, or transverse-field, circulator has the advantage of very high power handling capacity, even at high frequencies. These circulators are manufactured only in waveguide configurations. A typical unit is shown in Figure 2.10.

We will explain the operation of the differential phase shift circulator with the aid of Figure 2.11. Two sections of waveguide (EG and FH) connect two hybrid junctions. The junctions are magic tees or quadrature directional couplers. The two sections of waveguide are loaded with ferrite, and the ferrites are biased by an external dc magnetic field. The ferrites are located such that a signal traversing the waveguide in one direction undergoes a phase shift different from a signal traveling in the opposite direction. The name differential phase shift comes from this different phase shift behavior of the waveguide sections.

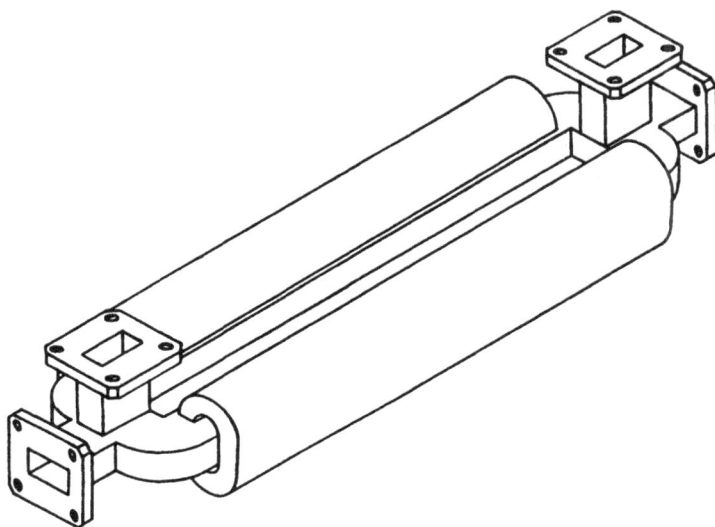

Fig. 2.10 A typical differential phase shift circulator.

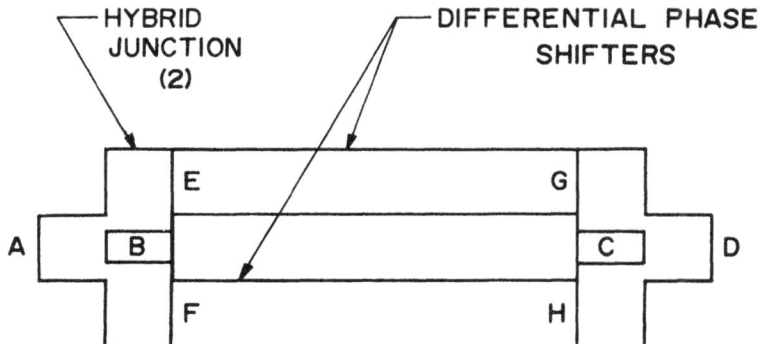

Fig. 2.11 The operation of the differential phase shift circulator.

A signal entering the circulator at port A is split by the hybrid and presented to the two waveguide sections at E and F. The two signals are phase shifted and arrive at G and H. Because of the phase relationship of the two signals, a summation occurs at one port of the output hybrid, D for example. At port C, the signals cancel each other, so no power is delivered to this port. A signal entering port D is split and phase shifted different amounts than is the signal that originated at port

A. The sum port for a signal entering port D is port B. In a similar manner, a signal incident on port B is coupled to C, and port C is coupled to A.

More information about differential phase shift circulators is presented in Section 5.3.

Differential phase shift circulators can be built to cover the common waveguide bands, and octave-bandwidth performance has been achieved with double-ridged waveguide units. For rectangular waveguide, the bandwidth is often somewhat more narrow than the waveguide bandwidth, as performance over the broader band is not good.

Most differential phase shift circulators have insertion loss between 0.5 and 1.0 dB. Many have lower loss, but usually the loss is slightly higher than that of a junction circulator.

Despite the higher loss, differential phase shift circulators can handle very high power levels because the power can be spread out over a large area of ferrite. The peak power capability is nearly as high as the capacity of the waveguide itself. If a differential phase shift circulator does not have sufficient power handling capability for an application, do not use a circulator.

The disadvantages of the differential phase shift circulator are its relatively large size and correspondingly heavy weight, and the effort that must be expended to construct it. These circulators are more complicated than junction circulators, and as such are more difficult to assemble and adjust.

2.5 FIELD-DISPLACEMENT ISOLATORS

Field-displacement isolators and resonance isolators are actually not circulators at all, but they are closely related devices, so we include some information about them.

Very wide bandwidths can be accommodated with coaxial field-displacement isolators: two octaves and wider. Waveguide units do not, of course, have this bandwidth. This design cannot be used to build circulators because the RF energy incident on the output port is dissipated in an internal resistive element.

The construction of a waveguide field-displacement isolator is shown in Figure 2.12, and Figure 2.13 illustrates a microstrip unit.

The ferrite material in a field-displacement isolator is positioned so that the concentrations of energy near the ferrite for signals traveling in forward and reverse directions are very different. In a waveguide device, this position is near a wall. Coaxial devices are implemented using a wide, asymmetrical strip transmission line in contact with ferrite. The energy concentrations are different on opposite sides of the line.

Both coaxial and waveguide field-displacement isolators utilize resistance cards, one example of which is a Mylar film metalized with a resistive coating. The

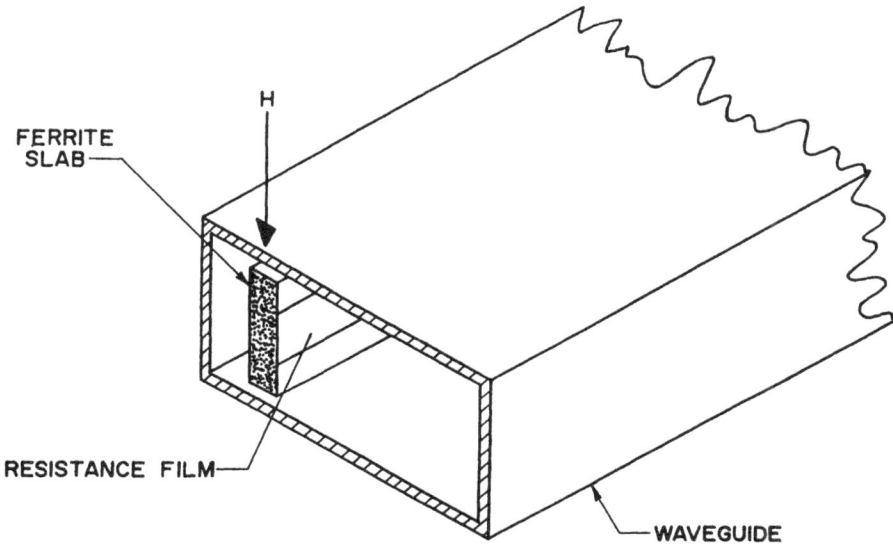

Fig. 2.12 Waveguide field-displacement isolator.

film in a waveguide isolator is placed near the ferrite so that it absorbs the energy concentrated in the area. The RF energy incident on the isolator output is absorbed in the resistance card. Energy entering the input does not become concentrated near the resistive material and is transmitted with low attenuation to the output. For reverse power absorption in a coaxial isolator, the resistance card is placed over one edge of the strip transmission line. The energy is concentrated at this edge for reverse-directed signals, and elsewhere for signals traveling in the forward direction.

Most practical field-displacement isolators covering broad bandwidths have insertion losses near 1.0 dB. It is possible to reduce the insertion loss to 0.2 dB or less if the bandwidth is narrow. It is difficult to obtain low insertion loss, because the RF energy cannot be kept out of the resistive material entirely over a broad range of frequencies.

The power handling of field-displacement isolators is limited because all the reverse power must be dissipated in the resistive film. Isolators with average power ratings of several hundred watts have been marketed.

Insofar as a wavelength-related variation of electrical energy is the basis of operation of field-displacement isolators, the operating frequencies are in the microwave range only. Units for lower frequencies would be prohibitively large.

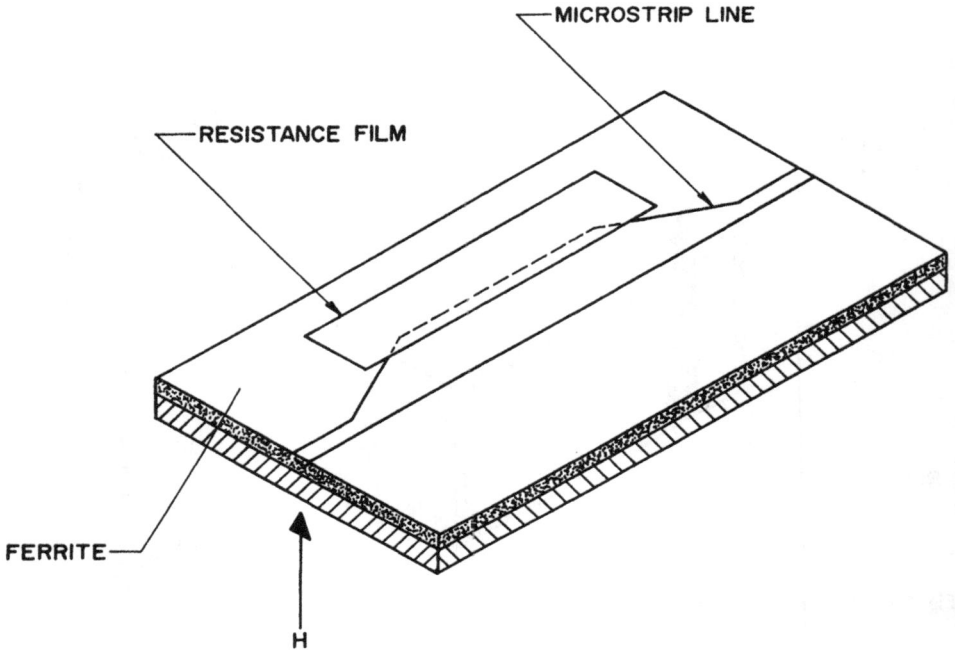

Fig. 2.13 Microstrip field-displacement isolator.

2.6 RESONANCE ISOLATORS

Another type of two-port ferrite device is the resonance isolator. Like the field-displacement isolator, this isolator does not have an accessible third port. The advantages of the resonance isolator are simplicity and a long, thin shape that can sometimes be fit into a tight place more easily.

Typical coaxial and waveguide resonance isolators are shown in Figure 2.14. The construction of a waveguide unit, shown in Figure 2.15, is very similar to its coaxial counterpart. There are several variations on the positioning and orientation of the ferrite, which will be discussed in more detail later.

The ferrite slabs in the waveguide are biased by a dc magnetic field supplied by an external magnet. The magnitude of the field is such that the ferrites are at ferrimagnetic resonance. The ferrites are positioned such that the electromagnetic field coupled into them is circularly polarized, and resonance absorption takes place for one direction of circular rotation as described in Chapter 1. A signal

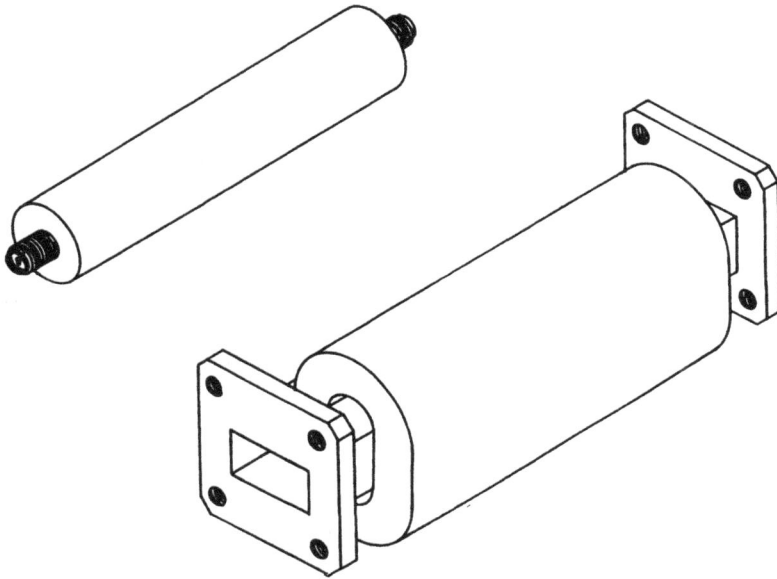

Fig. 2.14 Typical coaxial and waveguide resonance isolators.

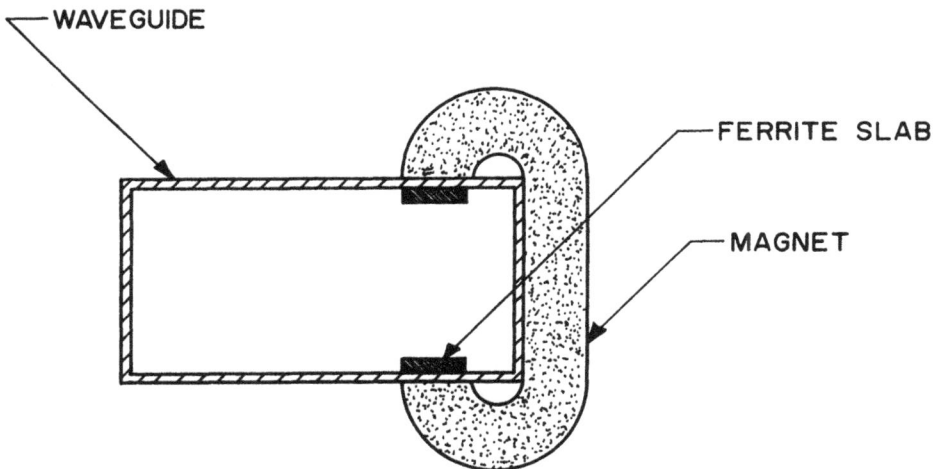

Fig. 2.15 Waveguide resonance isolator construction.

entering the isolator input propagates through the waveguide with little attenuation to the output. RF power traveling in the reverse direction suffers resonance absorption in the ferrites, so it does not appear at the input.

The reason for the difference in attenuations of the forward and reverse signals lies in the sense of circular polarization of the two waves in the ferrite. Forward waves traveling through the waveguide excite circularly polarized waves in the ferrite, while reverse waves excite circularly polarized waves with the opposite sense of polarization (the electric vectors of the two waves rotate in opposite directions). The rotation of the reverse circularly polarized mode coincides with the direction of electron precession, so the electromagnetic energy in the reverse wave is coupled into the precession. We say the reverse wave undergoes resonance absorption. Because the electrons can only precess in one direction for a particular polarity of applied dc magnetic field, and the rotation of the forward circularly polarized mode does not coincide with the direction of electron precession, the forward wave does not exhibit resonance absorption. More information regarding resonance isolators and how circular polarization is achieved is presented in Section 5.4.

Figure 2.16 is a cross-sectional view of a coaxial resonance isolator. The operation is the same as that of the waveguide isolator.

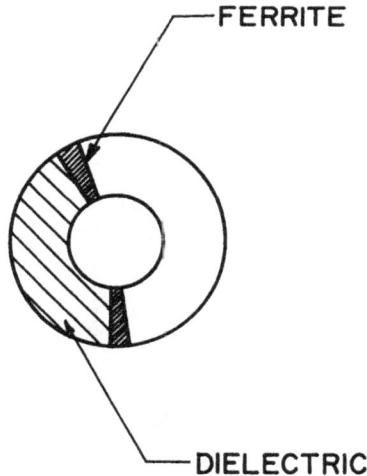

Fig. 2.16 Coaxial resonance isolator construction.

Coaxial resonance isolators can be built to cover octave bandwidths with approximately 1.0 dB loss and 10 dB isolation. We can obtain improved performance over narrower bands, and waveguide devices fall into this category.

The power-handling capacity of the coaxial variety of resonance isolator is on the order of several hundred watts of average power and 20–30 kW peak power. The ferrites in the waveguide isolator are easier to keep cool because the broad sides of the slabs are in contact with the waveguide walls. Therefore, the waveguide units have higher power capacities of tens of kilowatts average power.

REFERENCES

1. Collado R. O., and J. McCrea, "Learn to Specify High-Power Ferrite Circulators," *Microwaves & RF,* November 1987 and December 1987, pp. 107–114 and 91–97.
2. Soohoo, R. F., *Microwave Magnetics* (New York: Harper and Row, 1985).
3. Konishi, Y., "Lumped Element Y Circulator," *IEEE Transactions on Microwave Theory and Techniques,* November 1965, pp. 852–864.
4. Microwave Ferrimagnetic Circulator and Isolator Catalog (Kensington, MD: Cirqtel).
5. Military Specification MIL-C-28790, January 1973.
6. *Military Handbook* MIL-HDBK-216, April 1967.
7. Gartzke, D. G., "Determination of RF Peak and Average Power for Coaxial Connectors" (Port Salerno, FL: Solitron/Microwave, 1982).
8. General Connector Catalog (Oak Brook, IL: Amphenol, 1983).
9. Sekhon, W. S., R. G. Sanders, and C. Nugent, "Understanding Coaxial Circulators and Isolators," *Microwave Systems News,* June 1979.
10. *Military Handbook* MIL-HDBK-217D, January 1982.

Chapter 3
Applications of Circulators

3.1 LOAD ISOLATION

The principal application of the circulator is in providing isolation. When used as an isolator, the circulator has one port terminated in a matched load as described in Section 2.1. A two-port isolator is, of course, already in the proper configuration for providing isolation.

When we speak of isolation, we are speaking of a property whereby there is low transmission loss in one direction and relatively high loss in the other (reverse) direction. This isolation can be achieved using one device or many in series, and may be large in magnitude or small. For our present discussion, we will consider neither the magnitude of the isolation nor how it is obtained. We will represent the isolator by the symbol in Figure 2.2.

There are four important types [1] of vacuum tubes used to generate or amplify microwaves. These are the klystrons, magnetrons, TWTs (traveling wave tubes), and BWOs (backward wave oscillators). Solid-state devices in the form of diodes or transistors serve as oscillators or amplifiers as well.

All types of microwave oscillators can be subject to frequency shifting due to variations in load impedance (load pulling). If we install an isolator between the oscillator and the load, as shown in Figure 3.1, the oscillator can still deliver power to the load, but reflections from the load are attenuated in the isolator before they can return to the oscillator. Thus, the oscillator sees an essentially constant impedance. An isolator can take the place of a buffer amplifier in many applications, reducing power consumption and, in some cases, cost.

Amplifiers are not always unconditionally stable [2]. If an amplifier does not see the correct load impedance, it could break into oscillation at one or several frequencies, near or far away from the desired signal frequency. Not only are

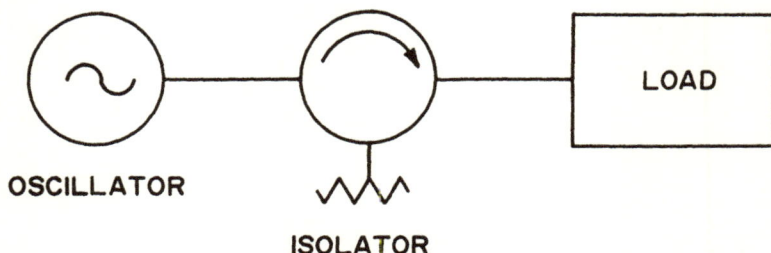

Fig. 3.1 Oscillator isolation.

these oscillations undesirable, but they could be damaging to an expensive amplifier. Isolators find application in isolating amplifiers from the loads they drive, even if these loads happen to be successive amplifiers in a chain, as shown in Figure 3.2.

Where an oscillator or amplifier drives a higher-power device, there is a possibility that the device could reflect back the signal from the amplifier or oscillator at a much higher amplitude, causing severe damage. Here again, an isolator rated to withstand the highest expected reflected signal level will protect the source.

In a receiver, an isolator may be placed in line with the antenna, as shown in Figure 3.3. The isolator may serve to prevent radiation of the local oscillator signal or to provide a good match for the transmission line coming from the antenna. The input VSWR of an LNA (low-noise amplifier) can be high because the input impedance for optimum low-noise performance may differ substantially from the transmission line characteristic impedance. Because of this, it may be desirable to use an isolator to improve the impedance match.

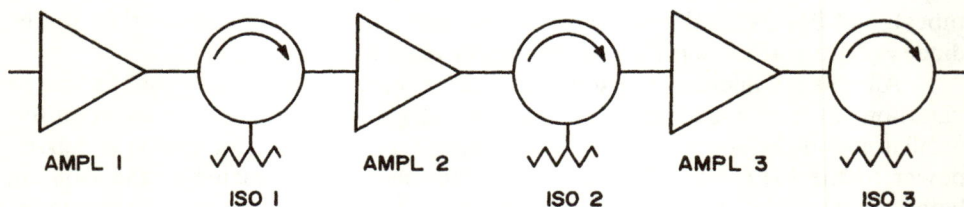

Fig. 3.2 Amplifier isolation.

3.2 DUPLEXING

A duplexer [3] is a device that switches an antenna to either a transmitter or a receiver, so that the same antenna can be used for both. There are several types of

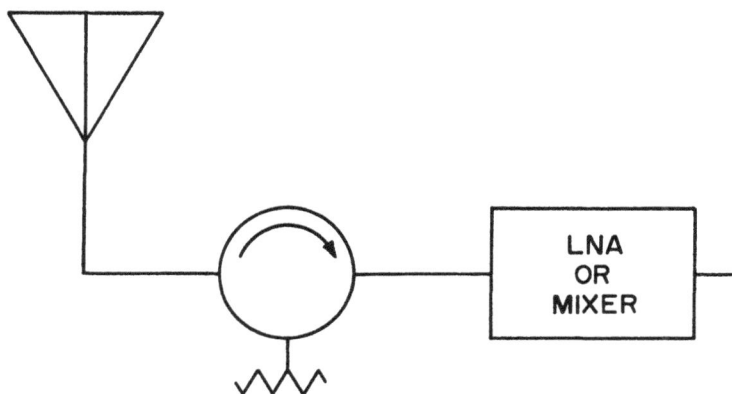

Fig. 3.3 Receiver application.

duplexers, including gas-discharge devices, solid-state switches, hybrid junctions, and ferrite units.

A ferrite duplexer in its simplest form is a circulator, connected as in Figure 3.4. The transmitted signal is coupled to the antenna, with only reflections from the antenna due to impedance mismatch coupled to the receiver. Signals incident on the antenna are coupled to the receiver. This scheme has the advantages of circuit simplicity, no need for mechanical or electrical switching, and high reliability. Also, transmission and reception can occur simultaneously.

Where only a circulator is used for duplexing, the signal leaking from the transmitter to the receiver, or TR isolation, is attenuated by the return loss of the antenna seen by the circulator. Therefore, a poor antenna VSWR (low return loss) results in poor TR isolation. To improve the isolation, we can add a limiter or switch to the receiver port of the circulator, as shown in Figure 3.5(a). The limiter could be a solid-state device, a tube type of limiter, or a ferrite limiter. Both the tube and solid-state limiters have the disadvantages of spike leakage and finite recovery time. A switch could be triggered by a high level of return signal from the antenna or by a separate signal from a control circuit.

Although ferrite limiters and switches are beyond the scope of this book, we will briefly describe their operation here to improve our understanding of duplexer theory.

The limiters rely on the nonlinear properties of ferrites. When the RF power exceeds a certain threshold, the insertion loss of the limiter rapidly increases. This nonlinearity is undesirable for the construction of circulators but finds use in limiters. Spin waves are responsible for the nonlinearity of ferrite limiters.

There are several methods [4] of constructing ferrite limiters. One is to place a ferrite sphere, magnetically biased at resonance, between two orthogonal strip-

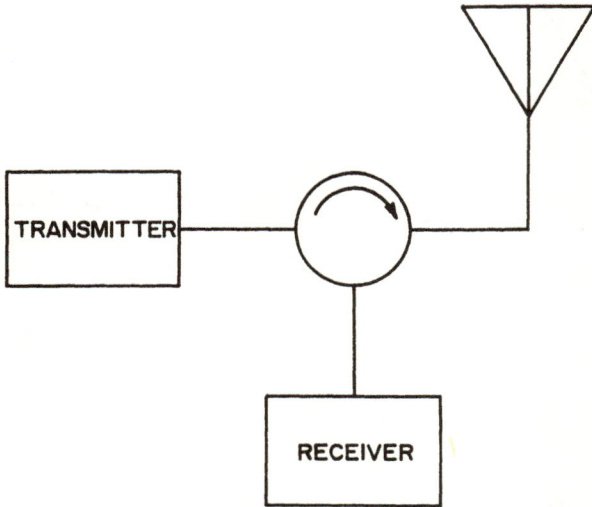

Fig. 3.4 A simple ferrite duplexer.

line resonators. This orthogonal stripline limiter is shown in Figure 3.5(b). The input signal is coupled to the sphere by one resonator, and the output is taken from the other. When the input signal level is high enough, the ferrite magnetic spin moment precesses sufficiently to prevent coupling the signal to the output resonator.

Other types of ferrite limiters include the cavity limiter, shown in Figure 3.5(c), where the ferrite sphere is located inside a resonant cavity; the comb limiter, shown in Figure 3.5(d) where a ferrite sphere is used to couple two or more quarter-wavelength resonators in a waveguide; and the subsidiary-resonance limiter, shown in Figure 3.5(e), in which a ferrite slab in a waveguide, operated in the region of the subsidiary resonance that occurs when the power threshold is exceeded, provides limiting.

Ferrite switching can be accomplished by reversing the magnetic field applied to a circulator, which reverses the direction of circulation, so that a signal incident on the input is no longer coupled to the output. Ferrite materials that retain their magnetization after the magnetizing field is removed find application in switches. With these materials, all that is needed to change the state of the switch is a low-energy current pulse.

We can synthesize a reciprocal ferrite switch [5] by connecting the circulator junctions, as shown in Figure 3.6. In the on state, the input is coupled to the output and vice versa, so power may be transmitted in either direction. In the off state, both input and output appear reflective.

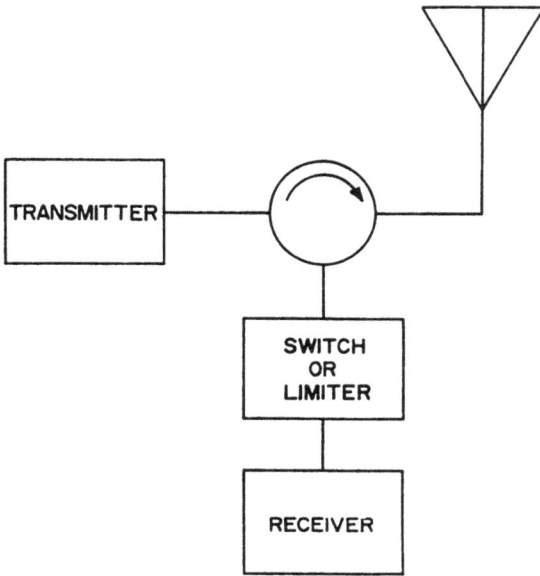

Fig. 3.5(a) A ferrite duplexer with improved transmit-receive isolation.

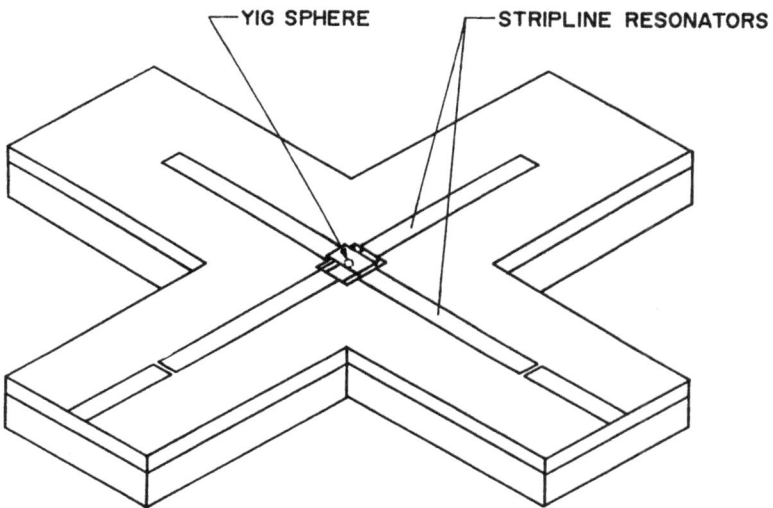

Fig. 3.5(b) Orthogonal stripline limiter.

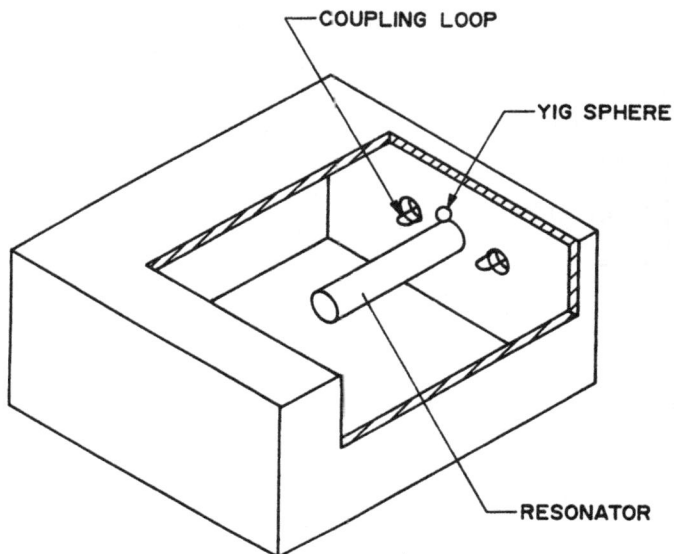

Fig. 3.5(c) Cavity limiter.

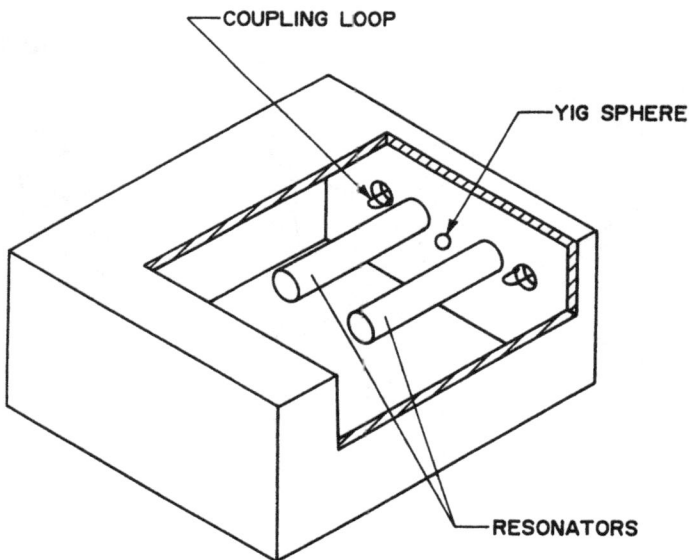

Fig. 3.5(d) Comb limiter.

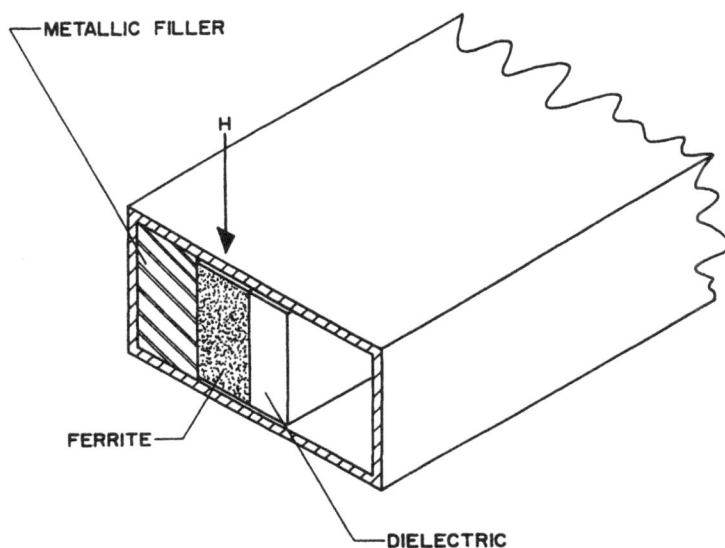

Fig. 3.5(e) Subsidiary-resonance limiter.

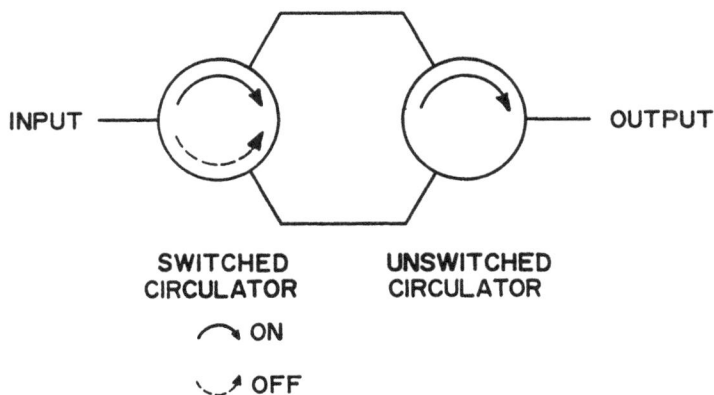

Fig. 3.6 The reciprocal ferrite switch.

3.3 MULTIPLEXING

A multiplexer [6] is a device composed of interconnected filters that can split a single channel carrying many frequencies into two or more channels carrying narrower bands of frequencies. The inverse process can also be carried out by a multiplexer; that is, combining two or more channels carrying different bands of frequencies into a single broadband channel.

A diplexer is a two-channel multiplexer. It is typically made from a low-pass filter and a high-pass filter, whereas a multiplexer having three or more channels is typically constructed using bandpass filters. A diplexer and a multiplexer, both based on circulators, are shown in Figures 3.7 and 3.8, respectively.

Diplexers and duplexers are frequently confused. One reason for this is the similarity in the names. Another is the fact that a diplexer could be used as a duplexer if the transmit and receive frequencies were different. A duplexer, however, is not a frequency-selective device like a diplexer.

We can build a multiplexer without a circulator by simply interconnecting the filters, but the filters must be specially designed to avoid interaction between them. If we use a circulator, we can build the multiplexer with almost any type of filter. In some cases, the filters are part of the equipment connected to the multiplexer, so the multiplexer consists of only a circulator.

A typical application of a multiplexer is to connect more than one transmitter or more than one receiver to a common antenna, which eliminates the necessity of having more than one antenna.

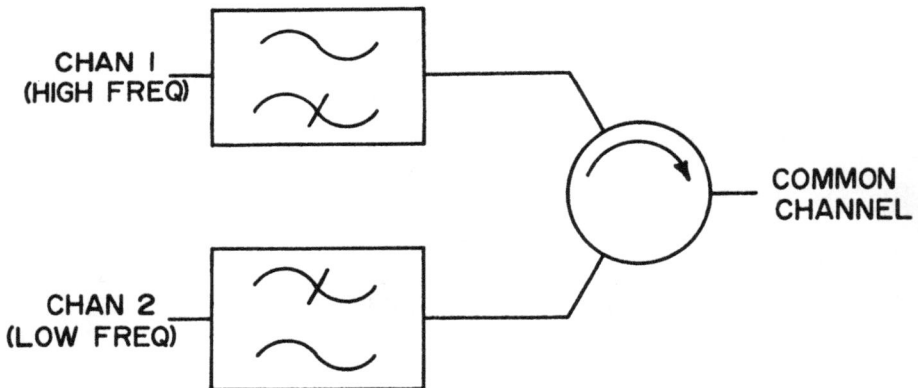

Fig. 3.7 Diplexer.

The between-channel isolation of a multiplexer depends on the filter rejection at the frequency of interest. Multiplexers that combine channels close together in frequency necessarily contain filters having very sharp skirts (rapid increase in rejection with frequency outside the passband).

To minimize multiplexer insertion loss, we need to be careful how we connect the circulator, because a signal entering the common port of the multiplexer may be reflected off a number of filters and pass through a number of circulator

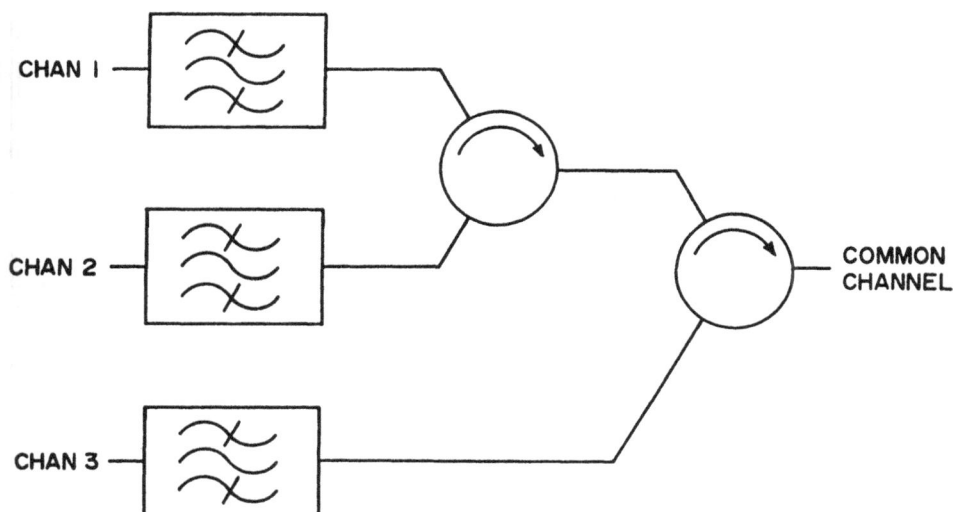

Fig. 3.8 A three-channel multiplexer.

junctions before arriving at its intended channel. Insofar as higher frequencies are attenuated more than lower frequencies in a coaxial medium, it would be wise to arrange the multiplexer so that higher frequencies do not have to travel as far as lower frequencies. If a particular channel must have lower insertion loss than other channels, it should be positioned first in line.

3.4 PARAMETRIC AMPLIFIERS

A parametric amplifier achieves signal gain, with or without frequency conversion, using a nonlinear, time-varying reactance [7]. Many *parametric amplifiers,* or PARAMPS, are built around circulators, so it is important for the ferrite device designer to know a little about these amplifiers and how they function.

The basic configuration of one type of parametric amplifier is shown in Figure 3.9. This PARAMP is called a nondegenerate one-port. The term nondegenerate refers to the relation of the pump frequency to the signal frequency. A degenerate PARAMP uses a pump frequency of twice the signal frequency. The pump source provides the energy that varies the reactance in the PARAMP to provide amplification. The power added to the input signal in the amplifier comes from the pump source.

There are various types of PARAMPs, including two-port, the one-port already mentioned, multiple pumps, degenerate, and nondegenerate. The one-port PARAMP with circulator is the most widely used type because of its good performance and circuit simplicity.

We can represent the PARAMP block in Figure 3.9 by the schematic of Figure 3.10. This circuit contains three resonators: one tuned to the signal frequency, one for the pump, and an idler, tuned to the difference between the signal and pump frequencies. The pump source drives the varactor diode so that its capacitance changes periodically at the pump frequency. The input signal, entering the PARAMP proper by way of the circulator, is also incident on the varactor. The pump and signal frequencies mix in the nonlinear varactor, and the desired mixing product is picked out by the resonant idler. The idler current remixes with the pump current in the varactor to produce a strengthened version of the input signal. This signal, which is in-phase with the original input signal, is reflected back toward the circulator where it is coupled to the amplifier output.

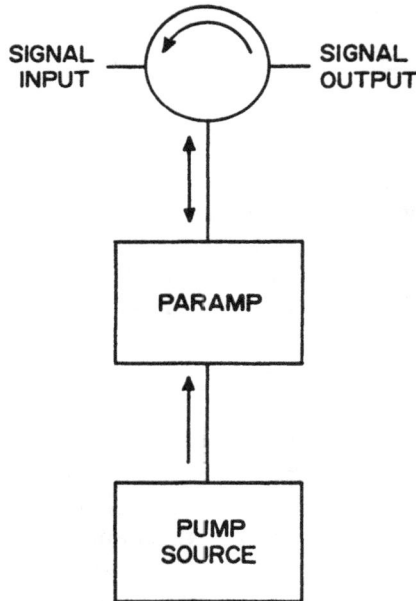

Fig. 3.9 The parametric amplifier.

The power flowing out of the pump source is delivered to both the signal port and the idler. Because the power flows out through the same port where the input signal enters, we need a circulator to separate the input and output ports of the amplifier.

Parametric amplifiers usually operate at relatively low power levels, so circulator power handling is generally not a consideration. However, low insertion

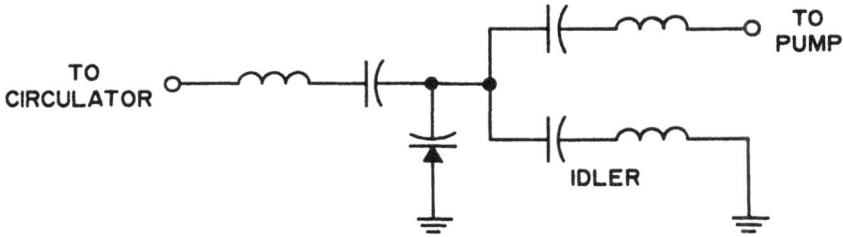

Fig. 3.10 PARAMP schematic.

loss is desirable because PARAMP designers often need to achieve low noise figures. Provisions for mounting the amplifier diode and other circuitry may be included in the circulator design.

Parametric amplifiers that have no separate pump source can be built. An avalanche diode can be used to oscillate at one frequency, supplying the pump signal, and to amplify at another frequency.

Another type of amplifier, very similar to the parametric amplifier, is the one-port negative-resistance amplifier. This amplifier also must utilize a circulator to separate the input and output ports of the amplifier. The basic circuit for the negative-resistance amplifier is shown in Figure 3.11.

Some of the most popular diodes for negative-resistance amplifiers include the IMPATT (impact ionization avalanche transit time), TRAPATT (trapped plasma avalanche triggered transit), and BARITT (barrier injected transit time).

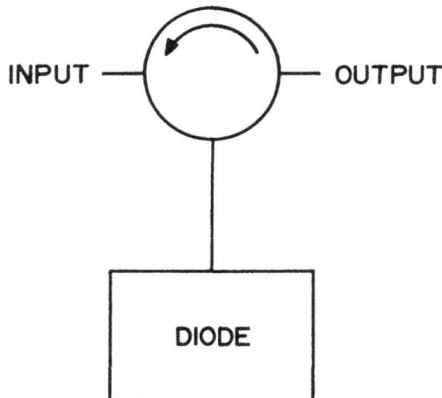

Fig. 3.11 The basic negative-resistance amplifier.

Insofar as the negative-resistance amplifier depends on the reflective properties of the diode (the input signal is incident on the diode and the output signal is what is reflected from the diode), the gain of these amplifiers is given by

$$\text{Gain} = \rho^2 = \left(\frac{Z_d - Z_0}{Z_d + Z_0}\right)^2 \tag{3.1}$$

where Z_d is the impedance of the diode. We see that impedance matching is very important to achieve optimum performance from a negative-resistance amplifier. In some applications, the impedance matching circuit may be included as part of the circulator.

Greater isolation between output and input for both the PARAMP and the reflective (negative-resistance) amplifier can be obtained with the configuration shown in Figure 3.12. We can include both circulator junctions in a single package, producing a four-port circulator or circulator-isolator combination.

The BARITT diodes offer lower noise performance than the IMPATT diodes, but the IMPATTs provide better power output. We can obtain high peak power capability and better efficiency from TRAPATT diodes, with greater difficulty in impedance matching.

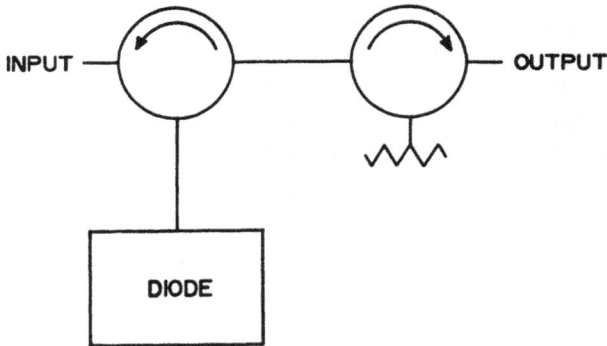

Fig. 3.12 A negative-resistance amplifier configuration to obtain higher output-input isolation.

Yet another type of amplifier that may use a circulator to separate input and output signals is the *maser*. The maser (microwave amplification by stimulated emission of radiation) is a predecessor [8] of the *laser,* which is now very much a part of our everyday lives. The maser is less common.

Masers amplify without the use of tubes, transistors, or diodes. Amplification is performed by raising the energy level of atoms or molecules in a solid or gas and then allowing the energy to be emitted. The energy level is raised by pumping with either a microwave signal or light.

One application of the maser is as a frequency standard. The energy-level transitions of the solid or gas cause emission of specific microwave frequencies that are very stable because they are frequency-dependent on the invariant atomic properties of the material.

A Maser amplifier making use of a circulator functions in much the same manner as the previously described parametric and negative-resistance amplifiers.

3.5 PHASE SHIFTING

Microwave circulators can be used in the construction of narrow-band phase shifters. The connection scheme is shown in Figure 3.13.

The reflective phase shifter block in Figure 3.13 may represent a varactor diode, a mechanical sliding short, or a section of transmission line shorted at various points along its length by PIN diodes. A fixed short circuit might also be placed here, and the magnetic field applied to the circulator junction could be varied to change the transmission phase shift.

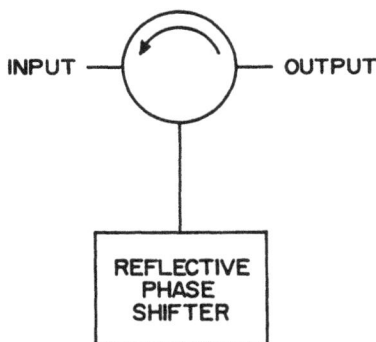

INPUT —— OUTPUT

REFLECTIVE
PHASE
SHIFTER

Fig. 3.13 Phase shifter.

The advantage of this type of phase shifter is simplicity; only a reflective phase shifter is needed instead of a transmission type. The circulator converts the one-port phaser to a two-port transmission device. Among the disadvantages is the frequency sensitivity. A constant phase shift cannot be maintained over a wide bandwidth with this design.

Several other types of ferrite phase shifters, not based on the circulator, exist. These include [9] toroidal waveguide, dual-mode, Reggia-Spencer, TEM, and Fox phase shifters.

The toroidal waveguide phase shifter, shown in Figure 3.14, uses rectangular ferrite toroids placed inside a section of waveguide with their axes coincidental with the waveguide axis. A wire is threaded through the centers of the toroids, and by application of a current pulse, a toroid can be magnetized with the desired magnitude and direction of magnetic field. Thus, the phase shift is changed by the current pulses. A phase shifter may contain one long toroid, and the magnetization is varied to change the phase shift; or several shorter toroids may be used, and the polarity of the magnetization is changed to switch phase shift sections.

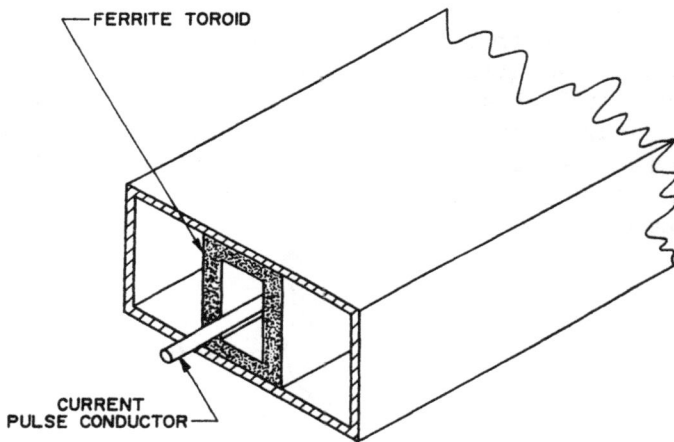

Fig. 3.14 Toroidal waveguide phase shifter.

The dual-mode phase shifter is akin to the Faraday rotation circulator. The end transitions are set up so that the energy is transmitted through the unit regardless of the direction of propagation, but different modes are used for the two phase states. The two modes use different polarizations in the phase-shifting ferrite rod: left-hand circular or right-hand circular. Figure 3.15 illustrates the scheme used in this device.

A Reggia-Spencer phase shifter consists of a ferrite rod or slab centered inside a section of waveguide, as shown in Figure 3.16. A solenoid around the waveguide provides the longitudinal magnetic field. The phase shift is varied by changing the magnetic bias.

The TEM phase shifters are coaxial versions that can be built using ferrite toroids or slabs in contact with a strip transmission line. Both types are depicted in Figure 3.17. The construction of a toroidal TEM phaser is very similar to that of the toroidal waveguide unit.

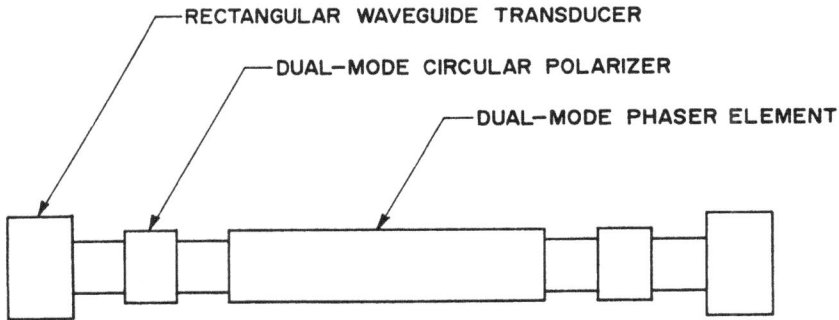

Fig. 3.15 Dual-mode phase shifter scheme.

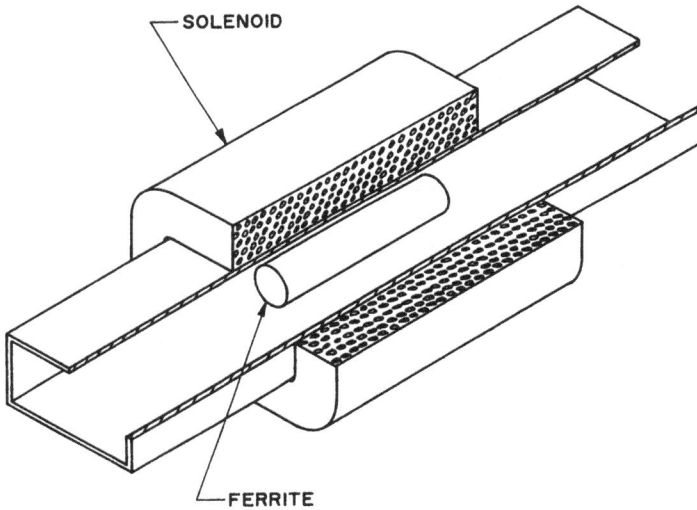

Fig. 3.16 Reggia-Spencer phase shifter.

The Fox phase shifter has transitions to achieve circular polarization in the center section. A half-wavelength ferrite plate is placed in this center section and is mechanically rotated to change the phase shift. This phase-shifting scheme is illustrated in Figure 3.18.

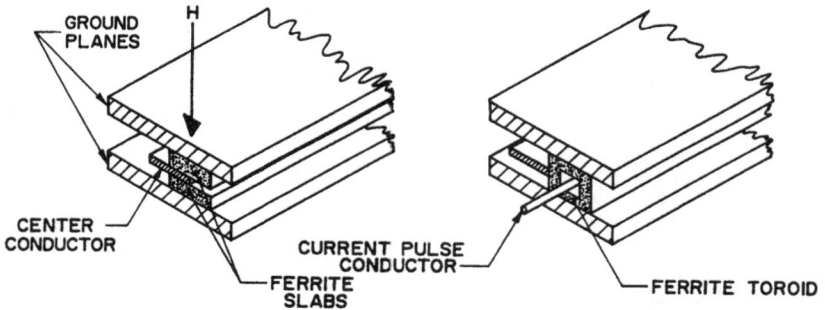

Fig. 3.17 Stripline TEM phase shifters: slab type (left) and toroidal type (right).

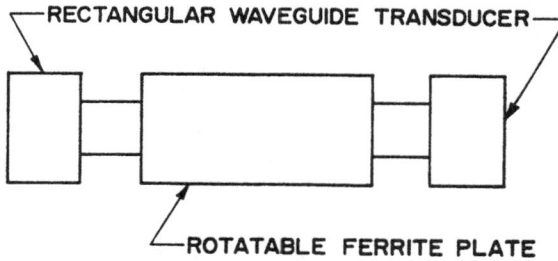

Fig. 3.18 Fox phase shifter scheme.

REFERENCES

1. Shrader, R. L., *Electronic Communication* (New York: McGraw-Hill, 1980).
2. Vendelin, G. D., *Design of Amplifiers and Oscillators by the S-Parameter Method* (New York: John Wiley and Sons, 1982).
3. Skolnik, M. I., *Introduction to Radar Systems*, 2nd Ed. (New York: McGraw-Hill, 1980).
4. Arams, F. R., M. Grace, and S. Okwit, "Low-Level Garnet Limiters," *Proceedings of IRE*, August 1961, pp. 1308–1313.
5. Clavin, A., "Reciprocal and Nonreciprocal Switches Utilizing Ferrite Junction Circulators," *IEEE Transactions on Microwave Theory and Techniques*, May 1963, pp. 217–218.
6. Matthaei, G. L., L. Young, and E. M. T. Jones, *Microwave Filters, Impedance-Matching Networks, and Coupling Structures* (Dedham, MA: Artech House, 1980).
7. Fink, D. G., and D. Christiansen, *Electronics Engineers' Handbook* (New York: McGraw-Hill, 1982).
8. Seippel, R. G., *Optoelectronics* (Reston, VA: Reston Publishing, 1981).
9. Whicker, L. R., *Ferrite Control Components*, Vol. 2 (Dedham, MA: Artech House, 1974).

Chapter 4
Material Selection

4.1 FERRITE SELECTION

The selection of ferrite material is perhaps the most critical aspect of circulator design. If the material is chosen incorrectly, the circulator will not perform as desired. In addition, if the material must be reordered from a vendor, a significant delay can result.

By the time the designer is ready to select materials, he or she will have some specifications for the circulator in question. The selection of ferrites depends on the operating frequency and bandwidth of the circulator as well as the RF power level, insertion loss desired, and the ambient temperature range. A good designer will also want to know what the material is and how it is made, so we will begin a description of ferrites and how they are manufactured.

Solid-state physicists like to name crystal structures after the minerals occurring in nature that have similar crystal structures. For this reason, the two general classes of microwave ferrite materials are called garnets and spinels. The mineral andradite (common garnet or black garnet) has the formula [1] $Ca_3Fe_2Si_3O_{12}$. If the silicon is replaced [2] by iron and the calcium by a rare earth such as yttrium, we have a new compound: the *yttrium iron garnet* (YIG).

The iron ions in the YIG structure are arranged in such a manner that some of their spin moments cancel each other and other spin moments are not canceled. This incomplete cancellation of spin moments is the basis of ferrimagnetism [3].

By substituting aluminum for some of the iron ions in the YIG, the net magnetic moment can be varied. Other rare earths such as gadolinium, holmium, or dysprosium may be added to change other properties such as spin-wave line width.

The mineral spinel has the formula $MgAl_2O_4$. Microwave ferrites (spinels) are similar in crystal structure, with a general formula of MFe_2O_4. The M may represent any one or a combination of the following ions: Al, Co, Li, Mg, Mn, Ni, Ti, and Zn.

Ferrites are manufactured using ceramic techniques. The exact procedures used vary among manufacturers and are usually secret. The raw materials used [4] include oxides, carbonates, oxalates, and nitrates.

The first step in the manufacture of ferrites is to grind or mill the raw materials to the correct particle size and then mix them together in the correct proportions. The mixing is often carried out in the presence of water or methylated spirit [2] to ensure homogeneity.

The mixture is then dried, if necessary, and "presintered" at a temperature somewhat lower than that used for final firing. The purpose of this firing is to cause a partial reaction between the raw materials and to help control shrinkage in the final firing. After presintering, the material is remilled and mixed with a binder to make shaping the material easier. One example of a binder is distilled water [2].

The paste can be formed to the desired shape through processes [4] such as extruding, pressing, and casting. The shaped parts are heated gently at first to drive off the binder.

Next, the ferrites are fired at temperatures of between 1000° and 1500°C in a carefully controlled atmosphere. The firing process can last from four hours to seven days, depending on the particular material.

The ferrites produced by this process are polycrystalline; that is, they have a nonuniform orientation of the crystal lattice. The properties of the polycrystals are an integration of the properties of a single crystal and are usually acceptable. In some applications, however, single crystals are desirable. These single crystals can be produced by other methods, but their size is limited and they are quite expensive.

Most manufacturing facilities and laboratories do not have the equipment to make ferrites, but many have the necessary machinery to cut and grind them. Ferrites can be purchased "as fired" (unmachined) in various shapes and sizes. These ferrites can be machined to the desired dimensions and surface finish. Being able to machine ferrites is convenient, even if they are purchased already machined, because changes in dimension are sometimes required to achieve optimum performance from circulators.

Ferrites are brittle, hard, and have low thermal conductivity, which makes them difficult to machine. Diamond wheels are usually required [5] to grind these materials, but in some situations silicon carbide wheels will suffice. Water should be used as a coolant during all grinding operations. The surfaces of ferrites are typically ground using a surface grinder, and the same machine with a diamond cut-off wheel can be used to slice the ferrite edges. The fact that ferrites are magnetic makes it easy to hold them by using a magnetic chuck.

The ferrites should be thoroughly cleaned after machining, because any surface impurities can degrade electrical performance.

Dimensional tolerances of finished ferrite parts are very important. If the thickness of a disk used in a stripline configuration were too great, the result could be a fractured ferrite. If the thickness were too small, the space between the ground planes would not be fully filled with ferrite, and a degradation in electrical performance would result. A typical thickness tolerance, including parallelism, is ±0.0005 inches. For the diameter of a disk or the altitude of a triangle, ±0.005 inches might be a reasonable figure. Smaller parts, of course, will require tighter tolerances. Smooth surface finishes are generally desirable, and a typical ferrite will have a finish of 16 microinches or better.

Another important consideration in the selection of ferrite materials is grain size. In below-resonance circulators, where the peak power threshold can cause higher insertion loss at higher peak power levels, reduced grain size can increase the threshold by an order of magnitude [6]. Normally, grain size is 10–20 μm. If the grain size is reduced, down to a practical limit of about one micron, the wavelengths of the longer wavelength spin waves is equal to the grain size. When the spin-wave wavelength and the grain size are similar, the spin waves are effectively broken up and the peak power threshold increases. The disadvantage of smaller grain size is higher cost.

A question that the designer of a junction circulator must face is whether to use a triangular ferrite or a disk. Nearly equal electrical performance can be obtained from either one, although experiment [7] has shown that lower loss might be achieved using a triangular geometry.

It is certainly less expensive to manufacture small disks than small triangles—they can be sliced from a rod. When considering larger ferrites, the disk part would not be cut from a rod, due to the immense proportions required of the rod and the cut-off blade. It is more economical to slice a triangle from a rectangular bar or to grind the edges of a fired triangle than to grind the circumference of a large disk.

Some engineers think that the raw material cost is lower if a triangle is used rather than a disk (see Figure 4.1). The truth is that for most ferrites, the raw material cost (especially for small parts) is a small fraction of the part cost. For small quantities and small parts, labor and manufacturing yield are larger terms than raw materials.

One must draw a dividing line between disk and triangle: At what dimension should you change over to a triangle from a disk? It is more cost effective to use disks where small ferrites are required, and more cost effective to use triangles where large ferrites are required. Since "large" ferrites are not normally encountered in below-resonance designs, and above-resonance circulators are primarily for lower frequencies (meaning large ferrites), a good rule of thumb is to use disks for below resonance units and triangles for above-resonance units.

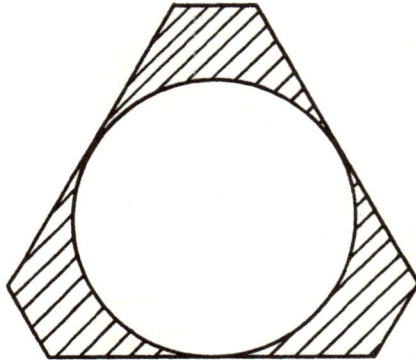

Fig. 4.1 Ferrite disk superimposed on electrically equivalent triangle. Shaded area is "wasted" ferrite.

Table 4.1 lists typical characteristics of commercially available garnets, and Table 4.2 the characteristics of spinels. The most important parameters of ferrite materials are the saturation magnetization $(4\pi M_s)$, line width (ΔH), Curie temperature (T_c), spin-wave line width (ΔH_k), and of lesser importance, the g-effective value, dielectric constant, and dielectric loss tangent.

The saturation magnetization is often measured [8] by the vibrating sample method. In this method, the test sample is a ferrite sphere approximately 2 mm in diameter. The spherical shape was chosen because in a sphere the internal magnetic field is more uniform than for any other shape. The ferrite sphere is placed in a strong dc magnetic field and vibrated to produce an induced voltage in a pickup coil. This induced voltage is proportional to the volume of the ferrite sphere and its saturation magnetization. If the voltage is compared to the voltage induced by a known reference sphere, the saturation magnetization can be calculated.

The test for resonance line width is made with a spherical sample inside a cavity resonator. The cavity is excited with an X-band RF signal, and a dc magnetic field applied to the sample under test is varied until maximum absorption occurs. If the RF frequency and the applied magnetic field are known, the effective gyromagnetic ratio, γ_{eff}, can be calculated from

$$\gamma_{eff} = \frac{\omega}{H} \tag{4.1}$$

Once γ_{eff} is known the g-effective value of the ferrite sample can be calculated from

$$g_{eff} = \gamma_{eff} \frac{2m_0 c}{e} \tag{4.2}$$

Table 4.1 Garnets

Composition	$4\pi M_s$	T_c	ΔH	ΔH_k	Composition	$4\pi M_s$	T_c	ΔH	ΔH_k
Y	1800 G	280°C	45 Oe	1.4 Oe	Y Gd Al Co	800 G	250°C	85 Oe	13 Oe
Y Al	175	90	40	1.5		1000	240	60	7.0
	250	100	40	1.5		1800	280	85	10
	300	115	30	2.0	Y Gd Al Dy	500	225	95	8.6
	350	125	40	1.5		600	175	85	8.9
	400	135	45	1.4		700	240	80	9.1
	550	160	40	1.4		800	245	70	3.0
	700	185	40	1.5		900	270	185	12
	800	200	40	1.5		1100	270	150	11
	1000	210	40	1.4		1200	260	60	9.4
	1200	230	40	1.4		1400	270	110	8.0
	1400	250	70	2.7		1600	280	75	6.6
	1600	265	40	1.4	Y Gd Al Ho	550	180	100	8.5
Y Gd Al	210	110	65	2.2		700	240	90	7.9
	400	150	65	4.2		800	240	110	8.1
	550	185	65	3.6	Y Gd Ho	1000	280	120	8.9
	700	200	60	4.0		1200	280	95	8.1
	800	220	75	5.2		1600	280	70	5.4
	800	260	55	4.3	Ca V In	600	200	25	1.2
	1000	250	75	3.6		800	205	10	1.2
	1200	260	50	3.2		1000	210	10	1.4
	1400	265	50	3.2		1200	220	10	1.2
Y Gd	720	280	200	7.6		1400	230	10	1.3
	900	280	140	6.4		1600	230	12	1.4
	1000	280	100	5.8		1850	240	15	1.3
	1200	280	75	4.3	Y Ca Zr	1950	240	60	
	1400	280	95	4.1					
	1600	280	50	3.8					

where m_0 is the mass of an electron, c is the velocity of light, and e is the unit electron charge.

The dc magnetic field applied to the ferrite sphere is varied until the resonance absorption drops to half power on each side of resonance (-3 dB points).

Table 4.2 Spinels

Composition	$4\pi M_s$	T_c	ΔH	ΔH_k	Composition	$4\pi M_s$	T_c	ΔH	ΔH_k
Mg Al	650 G	100°C	115 Oe		Ni Al	500 G	120°C	150 Oe	
	950	140	80			1000	400	320	
	1250	160	155			2100	560	460	6.1 Oe
	1700	225	120			2500	570	490	6.9
	2000	290	250	2.1 Oe	Ni Al Co	1400	425	260	20
	2420	310	180			1600	450	370	19
Mg Mn	1130	175	180	2.5		1800	500	775	25
	1400	210	260	2.0		1900	545	880	33
	1600	230	290	2.0	Ni Co	3000	585	350	12
	1900	280	350	2.0	Ni Zn	4000	500	270	
	2150	320	540	2.5		5000	375	160	
	2400	300	300	3.0	Li Ti	1000	330	300	1.5
	2800	300	300	2.9		1200	390	375	1.5
Mg Mn Al	750	90	120	5.2		2000	490	400	1.5
	1000	100	100	3.2		2200	500	450	1.5
	1300	140	135	2.6		2900	600	550	1.5
	1500	180	180	2.4	Li Ti Zn	1000	170	90	2.1
	1750	225	225	2.3		1300	210	150	2.1
Mg Mn Zn	2500	275	520	3.0		3000	375	150	2.1
	2800	225	540	2.3					
	3000	240	190	3.2					

The difference in magnetic field between these two 3 dB points is called the resonance line width.

The Curie temperature is determined by measuring the saturation magnetization as a function of temperature. The temperature where the saturation magnetization drops to zero is the Curie temperature. Only above this temperature is the ferrite material paramagnetic.

To measure spin-wave line width, the same ferrite sphere used for the other parameter measurements is placed at a point of maximum microwave magnetic field in a resonant cavity. The dc magnetic field is applied in the same direction as the RF field. This is called parallel pumping. The RF signal applied to the cavity is pulsed, and an RF output signal from the cavity is monitored. The dc magnetic field is varied until the output pulse starts to deteriorate, indicating the onset of nonlinearity in the ferrite sample. The microwave magnetic field, H_{RF}, can be calculated from the RF power applied to the cavity and certain constants of the cavity. From H_{RF}, the spin-wave line width can be calculated:

$$\Delta H_k = \frac{H_{RF}\gamma_{eff}4\pi M_S}{\omega} \tag{4.3}$$

The selection of ferrite material is substantially independent of the microwave transmission medium (waveguide, stripline, coaxial line), so the medium will not be considered in the following discussion. Neither will the sizing of ferrite parts, which will be covered in the following chapter on electrical design (Chapter 5).

The selection of an appropriate saturation magnetization is mainly dependent upon the microwave frequency, and three modes of operation will be considered: resonance, below resonance, and above resonance. The terms *above resonance* and *below resonance* refer to the magnetic operating point relative to ferrimagnetic resonance (see Figure 4.2). The magnetic operating points must not be confused with the operating points in the frequency domain, because they are just opposite—the below resonance region is higher in frequency than ferrimagnetic resonance. Low field loss will first appear at the upper frequency limit in a below-resonance circulator, and ferrimagnetic resonance will show itself at the upper frequency edge of an above-resonance circulator.

To avoid low field losses in both the resonance mode and the below-resonance mode, the saturation magnetization should be chosen such that the ferrite is fully biased magnetically. To meet this condition, we need [9]

$$4\pi M_s < \frac{\omega}{\gamma} - H_a \tag{4.4}$$

where H_a is the anisotropy field associated with the particular material selected. Typically, H_a is on the order of 100 Oe. Figure 4.3 is a plot of $4\pi M_s$ versus microwave frequency for $H_a \sim 100$ Oe, calculated using Equation (4.4). The maximum usable value of saturation magnetization varies linearly with frequency. For a resonance isolator, it is usually desirable to choose a high value of saturation magnetization to minimize the amount of magnetic field required. In the below-resonance mode, a lower value of saturation magnetization will increase [9] the peak power handling capability. The lower $4\pi M_s$ will also result in a narrower bandwidth for a given impedance-matching circuit complexity. One must decide which is more important—the bandwidth or the power handling. There are other methods of increasing the power handling that will be discussed later.

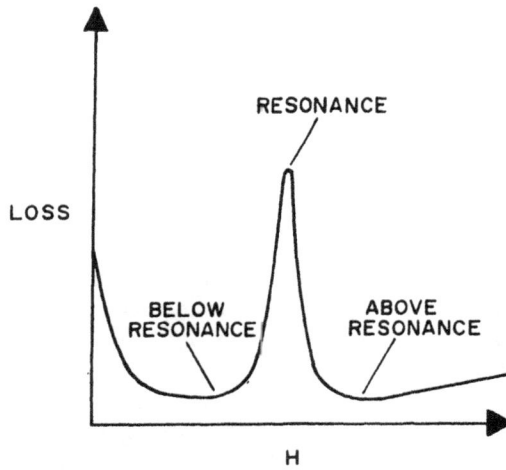

Fig. 4.2 Magnetic operating points: below resonance, resonance, above resonance. H = magnetization.

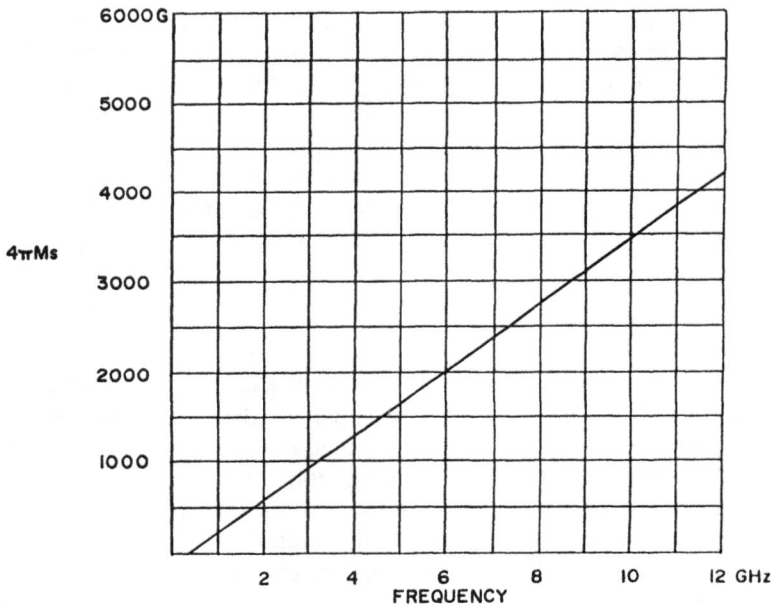

Fig. 4.3 Maximum values of saturation magnetization for below-resonance operation.

The saturation magnetization for above-resonance devices is not as critical as that for resonance and below-resonance devices. Because of the much higher magnetic fields, low field loss is not a concern. To achieve greater bandwidths, high values of $4\pi M_s$ should be selected. The mathematical relationship between saturation magnetization and bandwidth is presented in Section 5.1, but again, the choice of $4\pi M_s$ is not critical. At high frequencies, it may be necessary to select lower saturation magnetizations due to the extremely high magnetic fields required by higher saturation magnetizations and microwave frequencies.

The resonance line width for nonresonant devices (above- and below-resonance modes) should be as narrow as possible. This is especially important for the above-resonance mode. The narrow line width and moderate saturation magnetization of the calcium-vanadium garnets make them a good choice for broadband above-resonance devices where temperature stability is not important. In applications where high $4\pi M_s$ and narrow line width are needed, the lithium ferrites are desirable. YIGs and YAGs find usage where low $4\pi M_s$ and narrow line width are necessary. Excessive line width will result in increased insertion loss of a nonresonant device.

For a resonance isolator, we want to have a relatively broad resonance line width to achieve broadband operation. The bandwidth is proportional to the line width and the isolation (attenuation per unit length) is inversely proportional to the line width.

The ferrite material Curie temperature is a good indication of the material's temperature stability. For applications that require operation over a broad temperature range, it is important to consider the change in saturation magnetization with temperature. In many cases, high-power devices will also require temperature-stable materials because power losses are converted to heat. The Curie temperature is an indicator of temperature stability, as is the manner in which $4\pi M_s$ changes with temperature. Some materials have saturation magnetization characteristics that do not vary as much with temperature as do those of other materials. Figure 4.4 illustrates this concept graphically.

For an above-resonance circulator, we are not concerned with the nonlinear effects and subsidiary resonances that occur below resonance, so we select a material with a low spin-wave line width to avoid magnetic losses associated with a doped material.

Magnetic losses are caused largely by the ferrimagnetic resonance. If the ferrimagnetic resonance line width is higher, the resonance is broader, and resonance losses can manifest themselves in the operating frequency range of the circulator. Increased line width can be caused by either pores and other inclusions in the ferrite that reduce its density (higher density ferrites generally have lower

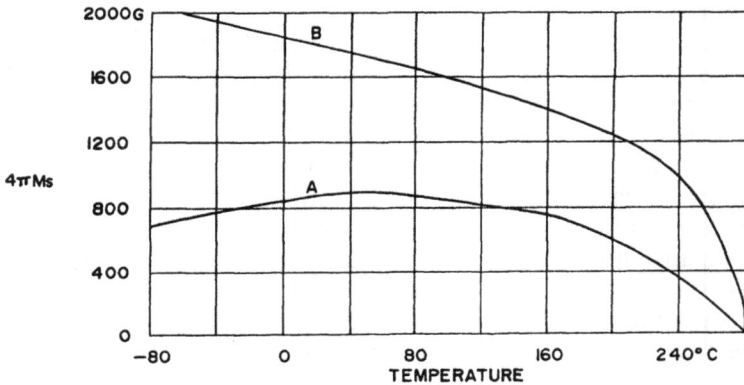

Fig. 4.4 Curie temperature is not everything. Materials A and B have identical Curie temperatures, but A is much more temperature stable than B.

line widths) or higher anisotropy due to random orientation of grains in the ferrite. Higher concentrations of rare-earth doping in garnets lead to higher anisotropy and hence higher resonance line width. Therefore, it is essential to hold rare-earth impurity levels low if narrow line width and low magnetic loss are desired. Higher doping levels are used in high spin-wave line width materials because the rare-earth ions tend to "pin" the spin waves.

Below-resonance devices built with low spin-wave line width ferrites can exhibit high power thresholds on the order of 100 W. For a YAG having a saturation magnetization of approximately 800 G, the threshold [10] can be increased by a factor of 30 by changing to gadolinium-doped YAGs. Y-Gd garnets would raise the threshold by a factor of 170. The trade-off here is increased insertion loss. In the case of the Y-Gd-Al garnets, the loss increases by a factor of 10 (at low power) and by a factor of 40 for the Y-Gd garnets. The high power threshold is proportional to the square of the spin-wave line width and inversely proportional to the saturation magnetization. A rough estimate of the increase in magnetic losses due to the use of a material with higher spin-wave line width can be obtained by comparing the squares of the resonance line widths.

Ferrites that have high g-effective values will require less magnetic bias than ferrites that have values near the theoretical value of 2.00. The converse is true of ferrites having low g-effective values—they require larger biasing magnets.

The dielectric loss tangent should always be as low as possible, but it is more important to keep dielectric losses low when the ferrite is positioned in a region of high electric field.

The dielectric constants of the garnets and spinels are relatively constant. For garnets, the figure is 14–16, and for spinels, 12–13.

4.2 MAGNET SELECTION

After ferrite materials have been selected and a preliminary electrical and magnetic design is developed, the designer will have some knowledge of the magnetic field requirements for the circulator, as well as the size or weight limitations imposed upon the design. The magnitude of the magnetic field together with the maximum size and weight of the magnetic components are the key factors in the selection of magnets.

Magnets commonly used in circulators fall into one of four broad classes: ceramic (barium ferrite), alnico (aluminum-nickel-cobalt), rare-earth cobalt, and neodymium-iron-boron. Table 4.3 lists some of the more important characteristics of these materials [11].

Table 4.3 Magnets

Material	Energy Product (MGOe)	Coercive Force (Oe)	Density (g/cc)	Max. Operating Temp. (°C)
Ceramic 1	1.00	1825	3.6	300
Ceramic 5	3.50	2200	4.9	300
Ceramic 8	4.30	2500	4.9	300
Alnico 5	5.50	640	7.3	540
Alnico 9	9.00	1300	7.3	540
Rare-earth cobalt 16	16.0	7900	8.3	250
Rare-earth cobalt 20	20.0	8300	8.3	225
Nd-Fe-B MQ 1	8.0	5300	6.0	125
Nd-Fe-B MQ 2	13.0	6500	7.5	150
Nd-Fe-B MQ 3	32.0	10500	7.5	150

Nd-Fe-B magnets are a relatively new development [12], and the material is prepared using a technique called jet casting. In this process, molten neodymium alloy is sprayed onto a rotating drum, which rapidly cools the material and forms it into ribbons. The ribbons are then crushed and the particles undergo further processing depending on the grade of magnet. For MQ 1 magnets, the particles are annealed and mixed with epoxy resin. This results in an inexpensive product that is easily machined. The energy product of MQ 1 is slightly higher than that of the ceramic magnets, and for a given application, the prices are similar. MQ 2 magnets are formed by hot pressing the particles from the jet caster. These magnets have energy products that compete with those of the rare-earth cobalt magnets. Further processing of the MQ 2 material results in the anisotropic MQ 3, which has the highest energy product of the materials mentioned. MQ 3, being anisotropic, has the disadvantage that it can only be magnetized in one direction.

The energy product of a magnetic material is the maximum product of flux density (B) and magnetic intensity (H) taken from the material demagnetization curve. This provides a measure of the "strength" of the material, or the size of a magnet necessary to provide a particular magnetic field. The coercive force (H_c) is the demagnetizing force required to reduce the induction (flux density) in the material to zero. Magnets that have low coercive forces should be charged (magnetized) in the magnetic circuit to get the maximum energy from them, because removing them from the circuit usually causes a loss of magnetism. Alnico magnets fall into this category and are less desirable because of their low coercive forces.

The most cost-effective biasing fields are provided by the ceramic and MQ 1 materials. Standard-sized shapes such as disks and rectangular bars should be used whenever possible to avoid costly machining. When machining is necessary, one will find that the MQ 1 material, being epoxy-based rather than ceramic, is easier to machine. Triangles are normally not available as standard shapes, so their use should be avoided when possible.

For applications demanding high magnetic fields and minimum size and weight, the rare earth cobalt and MQ 3 materials are preferred. MQ 3 has a higher energy product than the rare-earth cobalt materials but is not yet as widely available and may be more expensive in some cases.

Another factor worthy of consideration is the maximum operating temperature of the magnet. MQ 1 has a fairly low maximum operating temperature of 125°C. The other Nd-Fe-B materials have similar temperatures. Maximum temperatures at least 100° higher apply to the other materials in Table 4.3. Above these maximum operating temperatures, there is an irreversible loss of magnetic flux.

Reversible temperature effects in magnets can be advantageous. Because ferrite $4\pi M_s$ usually drops with increasing temperature, it is desirable to have reduced magnetic flux at higher temperatures. The ceramic magnets have the highest negative reversible temperature coefficient of about 0.9%/°C [13].

In applications where the magnets will not be charged in the magnetic circuit and a magnet charger is not available, magnets should be ordered from the vendor "fully oriented" and the direction (plane) of magnetization should be specified.

4.3 MAGNETIC COMPENSATING MATERIAL SELECTION

To design a circulator with a wide operating temperature range, it is often necessary to temperature-compensate the magnetic circuit. Magnet manufacturers strive to produce magnets that provide a nearly constant flux regardless of temperature, but for most circulators we want a decrease in flux as temperature increases.

Special materials are available that will provide a reduction in magnetic flux as temperature increases. These materials can be either ferrites or nickel-steel alloys. The manufacturers of the materials provide data indicating the change in permeability versus temperature for a specific H (magnetizing force). Some typical data are presented in Figure 6.6 (Chapter 6).

Once the required magnetic flux density as a function of temperature is determined, either by calculation or testing, an appropriate compensating material can be selected by examining the manufacturer's data.

The metals are far easier to machine than the ferrites, but are not usually available in the small quantities needed for most circulator requirements. A minimum order from a steel mill can be quite a large quantity, but the ferrites, on the other hand, are available in small quantities.

4.4 DIELECTRIC SELECTION

Many circulator designs will require the use of dielectric materials. Dielectrics provide insulation to prevent voltage breakdown and reduce the size of distributed capacitances (transmission lines) by increasing the electric flux density.

Dielectrics can be classified according to physical characteristics, dielectric constant, dielectric loss tangent, and dielectric strength.

The lowest loss dielectric is air, which has a relative dielectric constant of 1. Air does not have high dielectric strength, but the voltage breakdown characteristic can be improved by pressurization. Other gases, such as sulfur hexafluoride, are used where high dielectric strength is needed without sacrificing the low loss aspect of a gas dielectric.

Higher in loss and dielectric constant, we have silicone materials: greases, two-part liquids that cure to a rubberlike consistency, and self-curing silicone sealants available in various viscosities. These materials have a dielectric constant of about 2.7 and very good dielectric strength.

Rexolite and Teflon are two low-loss plastics that have dielectric constants of 2.5 and 2.1, respectively. Rexolite has a maximum temperature rating of only

100°C but has lower loss than Teflon, which is a much higher-temperature material.

Various printed-circuit laminates find use in circulators. These materials are available in a wide range of dielectric constants and thicknesses, and are usually copper-clad. Circulator construction can be simplified by using these laminates because the circuit can be etched onto the dielectric.

Composite dielectric materials consisting of high dielectric constant ceramics mixed with polystyrene are offered in various values of dielectric constant from 3 to 30. These materials, when cut into small blocks, are useful for tuning low-power circulators.

Ceramic dielectrics, made from oxides of aluminum, calcium, magnesium, and titanium, are usually marketed by ferrite manufacturers. Dielectric constants from 6 to over 100 are available, the dielectric loss tangent increasing with dielectric constant.

Ferrites can be purchased with ceramic dielectrics bonded to them. These assemblies make junction circulator construction simple, because the impedance transformers typically extend radially from the ferrite.

To select a dielectric material, it is wise first to determine the required dielectric constant. Physical size limitations placed on distributed capacitances or transmission lines require minimum values of dielectric constant. The upper limit of dielectric constant is set by the vanishingly small dimensions of the center conductor in a coaxial structure, the propagation of higher-order modes, the higher dielectric loss tangent, and the tighter dimensional tolerances required.

Once the required dielectric constant has been determined, the dielectric material can be selected by considering the expected voltage gradient in the dielectric and the dielectric loss tangents. The lowest-loss dielectric that has the required dielectric constant and dielectric strength should be chosen. Where further discrimination between materials is needed, the material that is easiest to form to the desired shape and size should be selected. The silicone materials (and air) are excellent in this respect.

4.5 METALS SELECTION

The metals commonly encountered in circulators are aluminum, brass, copper, silver, and steel. Characteristics to be considered in the selection of metals include cost, machinability, electrical conductivity, thermal conductivity, weight, and magnetic properties.

For a circulator housing, which usually provides ground planes, the metal selected should be inexpensive, easily machinable, lightweight, highly conductive, and nonmagnetic. Thermal conductivity is secondary in most cases. Aluminum is a good choice. The particular alloy chosen should be one that can easily be machined to a very smooth, flat surface. In applications where low insertion loss is specified, the housing can be plated with a more conductive metal, silver. The thickness of the plating should be such that most of the RF current flows in the silver. Five times the skin depth is a good rule of thumb for thickness. Water-cooled circulator housings may be made from copper, which has twice the thermal conductivity of aluminum.

Center conductors for coaxial circulators need to have very good electrical conductivity and easy machinability or etchability. In addition, the metal must be solderable for attachment to the RF connectors. Brass is a good choice because it is cheap, easily machinable, and available in many precise thicknesses as shim stock. Where the center conductor will be etched, copper is a better choice. If insertion loss is a critical specification, the center conductor may be silver-plated. This plating does not have a large influence over loss in most applications, however.

Cold-rolled steel is generally used for magnet pole pieces, magnetic shielding, and other magnetic circuit components. The steel will normally be plated for rust prevention.

Large waveguide circulators are typically constructed from commercially available waveguide tubing and flanges.

REFERENCES

1. *Handbook of Chemistry and Physics* (Cleveland: Chemical Rubber Publishing Co., 1939).
2. Clarricoats, P. J. B., *Microwave Ferrites* (New York: John Wiley and Sons, 1961).
3. Wert, C. A., and R. M. Thomson, *Physics of Solids* (New York: McGraw-Hill, 1970).
4. Publication No. 50030040 (Adamstown, MD: Trans-Tech, Inc., 1984).
5. *Tech-Briefs* (Adamstown, MD: Trans-Tech, 1973), p. 13.
6. *Tech-Briefs* (Adamstown, MD: Trans-Tech, 1973), p. 14.
7. Helszajn, J. and D. S. James, "Planar Triangular Resonators with Magnetic Walls,"*IEEE Transactions on Microwave Theory and Techniques*, February 1978, pp. 95–100.
8. *Tech-Briefs* (Adamstown, MD: Trans-Tech, 1973), p. 20.
9. Dydyk, M. "Take Two Steps Toward Better Circulator Design," *Microwaves*, March 1979, pp. 53–62.
10. Green, J. J., and F. Sandy, "A Catalog of Low Power Loss Parameters and High Power Thresholds for Partially Magnetized Ferrites," *IEEE Transactions on Microwave Theory and Techniques*, June 1974, pp. 645–651.
11. Magnet Catalog P5A (Hicksville, NY: Permag Corp., 1986).
12. Carlisle, B. H., "Neodymium Challenges Ferrite Magnets," *Machine Design*, January 9, 1986, pp. 24–30.
13. *Factors Affecting Magnet Stability*, Manual No. 14 (Valparaiso, IN: Indiana General Co., 1978).

Chapter 5
Electrical Design

5.1 JUNCTION CIRCULATORS

After we have an understanding of the basic theory of operation of the circulator, the tensor permeability of ferrite, and have written a set of specifications for the circulator to be designed, we proceed to select a ferrite material using the criteria outlined in Chapter 4. The next steps in the design process are the sizing of the ferrite parts, selection of the correct position of the ferrites in the circulator, and synthesis of impedance-matching networks. For coaxial circulators, appropriate center conductor geometries must be chosen. The magnetic and mechanical design of the circulator are covered in Chapters 6 and 7.

We begin this chapter with a discussion of the propagation of microwave energy in the circulator junction [1]. The junction circulator is based on two counter-rotating wave components or modes. As we know from Chapter 2, a circulator is a device with ports arranged such that energy entering a port is coupled to an adjacent port, but not coupled to the other ports. With this in mind, we can analyze the rotating modes shown in Figure 5.1. A three-port circulator is depicted in this figure, but the analysis could be extended for any number of ports. Usually, however, junction circulators with more than three ports are made by simply cascading three-port circulators.

The two modes shown in Figure 5.1 must travel at different velocities for circulator action to occur. For coupling from port 1 to port 2, the clockwise $(-)$ and counterclockwise $(+)$ modes must differ in phase by $N2\pi$:

$$2\beta_- L - \beta_+ L = 2N\pi \qquad (5.1)$$

where β is the phase constant of the positive or negative mode. Note that the negative wave must travel twice the distance the positive wave travels to reach

---

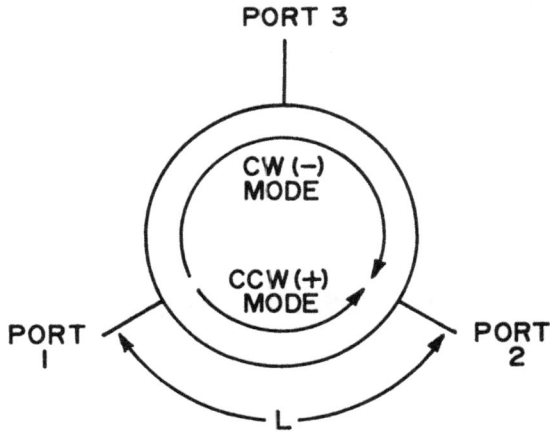

Fig. 5.1 Rotating modes in a three-port circulator.

port 2. N denotes any integer and L is the distance traversed by a wave. For port 3 to be decoupled or nulled, the two wave components arriving at this port must differ in phase by an odd multiple of π:

$$-\beta_- L + 2\beta_+ L = (2M - 1)\pi \tag{5.2}$$

where M is any integer. If we solve Equations (5.1) and (5.2) simultaneously, we derive

$$\beta_- L = \frac{4N + 2M - 1}{3} \pi \tag{5.3}$$

$$\beta_+ L = \frac{2N + 4M - 2}{3} \pi \tag{5.4}$$

When Equations (5.3) and (5.4) are satisfied, we have perfect circulation.

The two counter-rotating waves propagate in a ferrite disk with propagation constants given by Equations (1.37) and (1.38) when the disk is biased with a dc magnetic field parallel to its axis (transverse to the direction of propagation). The waves form standing-wave patterns as shown in Figures 5.2 and 5.3 for disks and triangles, respectively. Any ferrite shape that has threefold symmetry could be used, but the two simplest shapes are the disk and the triangle. These shapes are the ones used in nearly all commercial circulators.

The standing-wave pattern for an unmagnetized ferrite disk is shown in

ISOLATED
PORT

PORT 3

PORT I

INPUT

PORT 2

OUTPUT

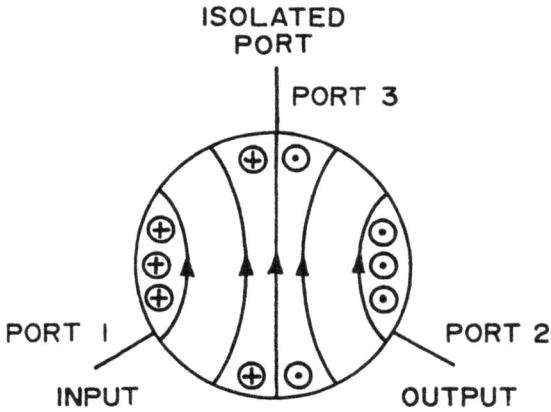

Fig. 5.2 Ferrite disk standing-wave pattern. RF electric field lines are perpendicular to the disk plane and the magnetic lines are parallel to the disk plane.

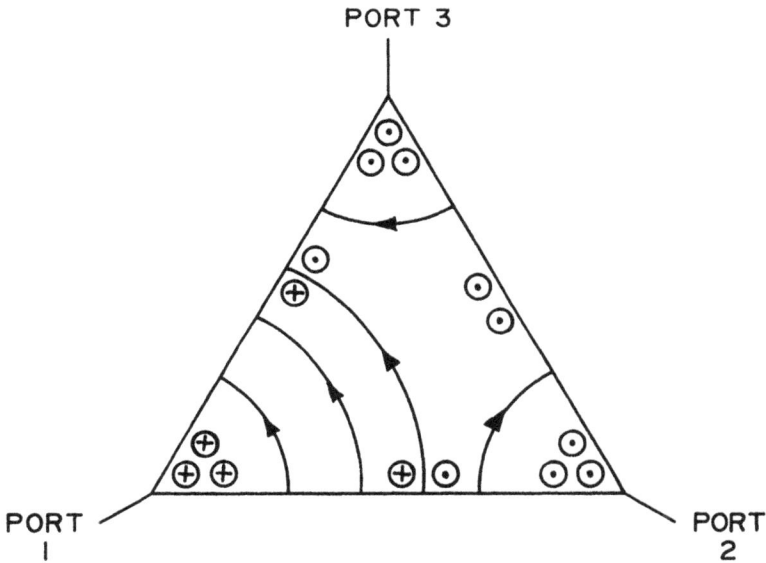

PORT 3

PORT
1

PORT
2

Fig. 5.3 Ferrite triangle standing-wave pattern.

Figure 5.4. Notice the difference between this pattern and the one for a magne-tized disk in Figure 5.2: it is slightly rotated. In an unmagnetized disk (or triangle) the phase and amplitude of the signals appearing at ports 2 and 3 are equal and,

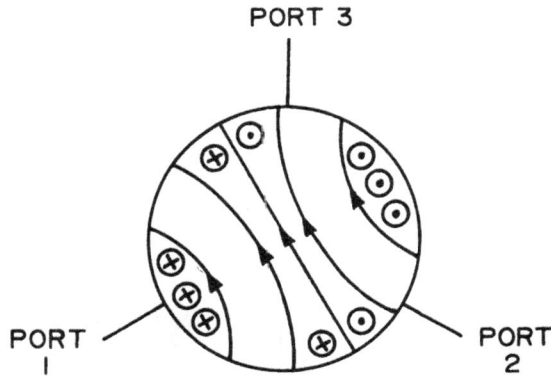

Fig. 5.4 Unmagnetized ferrite disk standing-wave pattern.

ideally, the VSWR at port 1 is 2:1. The two rotating modes are said to be degenerate; the disk resonances due to them are at the same frequency. When a dc magnetic field is applied to the ferrite, the propagation constants of the positive and negative rotating modes are no longer equal, and the resonant frequencies for the two modes, ω_+ and ω_-, are different. The standing-wave pattern rotates, and the coupling to ports 2 and 3 from port 1 is not the same. The operating frequency of the circulator is between the disk resonant frequencies due to the two counter-rotating modes.

The factor κ/μ from Equation (1.39) not only relates orthogonal microwave magnetic field components, but describes the amount of frequency splitting between the two resonance frequencies for the counter-rotating modes. It also determines the amount of rotation of the standing-wave pattern and the anisotropy of the ferrite; κ and μ are elements of the Polder permeability tensor, defined by Equations (1.30) to (1.33).

Now that we have shown that the ferrite disks or triangles are resonators, we need an algorithm for determining the size of these resonators from the circulator specifications. Engineers with experience in the design of circulators usually have repertoires of geometries in their memories upon which they can call when asked. Unfortunately, someone with no experience in circulator design has no idea where to start. We could print tables of ferrite dimensions in a book, but normally the needed information would have to be interpolated from the tables. Now, we need to know how to interpolate and arrive at a usable design. It is also helpful to learn the effects of changes in the various circulation parameters. For these reasons, and because most circulator manufacturers are unwilling to part with their valuable design data, we will not present tables.

A complicating factor in the presentation of specific circulator design information is that no universal equations exist that will, in all cases, produce working circulators. Much of circulator design [2] has "proceeded on an empirical basis." This is not to say that we should not look at the equations that have been derived, because they do have value in that they can teach the relationships between the various circulator parameters. In addition, some concepts, such as the permeability tensor, are difficult if not impossible to visualize without the use of mathematics.

To begin our development of a ferrite-sizing algorithm, we will briefly describe some papers by early workers in the circulator industry: Bosma [3], Fay and Comstock [4], Helszajn [5,6,7], Davies [8], Auld [9], Simon [10] and Green and Sandy [11]. We will not present all the mathematical details from these papers, only some results. All of these papers are required reading for the circulator expert.

The papers include approximations and assumptions to simplify the mathematics (which is one reason no universal, exact design equations exist), some of which include the following:

1. The ferrites are usually considered loss-less and magnetically saturated to avoid low-field losses.
2. Fringing at the edges of striplines is ignored.
3. The field intensities do not vary over the width of stripline conductors.
4. Striplines are purely in the TEM mode.
5. There is no z-coordinate variation of electromagnetic fields in the ferrites.
6. Coaxial center conductors are between two ferrites (the fields are the same on both sides of the conductor).
7. The electromagnetic fields fall off immediately at the ferrite and center conductor edges.

The paper by Bosma [3] is concerned with the stripline above-resonance circulator. He states that the magnetic bias applied to this type of circulator is more than four times greater than the field required for resonance in most cases. Bosma uses a scattering matrix approach to circulator synthesis. A scattering matrix is a matrix (3-by-3 in the case of a circulator) of reflection coefficients and transmission coefficients that describe the circulator junction. The matrix lends itself to devices that are symmetrical. The ferrite disks and center conductor are assumed to have the same diameters, and are arranged as shown in Figure 2.8. Figure 5.5 indicates the key dimensions of the stripline center conductor. These dimensions are related by

$$\sin \psi = \frac{W}{2R} \tag{5.5}$$

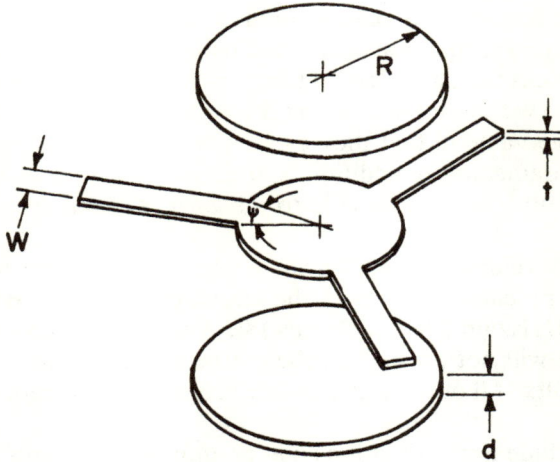

Fig. 5.5 Stripline center conductor and ferrite dimensions.

The effective relative permeability of the magnetized ferrite, a factor in Equation (1.37), is given by

$$\mu_{\text{eff}} = \frac{\mu^2 - \kappa^2}{\mu} \qquad (5.6)$$

Bosma also introduces a wave number:

$$k^2 = \omega^2 \mu_0 \varepsilon_0 \mu_{\text{eff}} \varepsilon \qquad (5.7)$$

He presents equations for the electric field, parallel to the z-axis (ferrite disk axis), and the magnetic fields, tangential and radial components, in the ferrites. These equations are related by a Green's function, and expressions for the relations between the fields in the striplines and the fields in the ferrites are derived. The Green's function can be thought of as a transfer impedance function for the electromagnetic fields in the ferrites.

The solutions of the electromagnetic field equations involve the Bessel function of the Nth order. For most practical circulators, $N = 1$ and for resonance we have

$$J_1'(kR) = 0 \qquad (5.8)$$

where J denotes the Bessel function, k is the wave number from Equation (5.7), and R is the ferrite disk radius. If we evaluate Equation (5.8), we find that $kR =$

1.84. This equation is valid when the ferrites are not magnetized, that is, when the rotating modes are degenerate. Because the resonance frequency from Equation (5.8) is between the two resonance frequencies for the two counter-rotating modes when the disks are magnetized, it is an approximation of the operating frequency of the circulator.

Bosma explains that $\omega \sqrt{\mu_0 \varepsilon_0} = 2\pi/\lambda$ and derives from (5.7) and (5.8):

$$R = \frac{1.84\lambda}{2\pi \sqrt{\mu_{\text{eff}}\varepsilon}} \tag{5.9}$$

where ε is the relative dielectric constant of the ferrite material. To obtain a simpler expression for μ_{eff} than Equation (5.6), he assumes that $\omega_0^2 \gg \omega^2$ because the ferrites are biased far above resonance and $\omega_0^2 \gg \alpha^2$ because the resonance losses need not be considered, and by manipulation of the Equations for μ and κ (Equations (1.30) to (1.33)), derives

$$\mu_{\text{eff}} = \mu = \frac{H_{\text{dc}} + M_0}{H_{\text{dc}}} \tag{5.10}$$

$$\kappa = \frac{M_0}{H_{\text{dc}}^2} \tag{5.11}$$

The imaginary components of μ and κ do not need to be considered because the ferrites are biased far above resonance.

Substituting Equation (5.10) into (5.9), we arrive at an expression for the ferrite disk radius as a function of wavelength, ferrite saturation magnetization, dielectric constant, and applied magnetic field:

$$R = \frac{1.84\lambda}{2\pi \sqrt{\varepsilon}} \sqrt{\frac{H_{\text{dc}}}{H_{\text{dc}} + 4\pi M_{\text{s}}}} \tag{5.12}$$

Bosma suggests restrictions on stripline width to maintain the magnetic field in the ferrite as high as possible, a desired operating point for above-resonance circulators:

$$W < \frac{\lambda}{30} \tag{5.13}$$

$$W < 0.75R \tag{5.14}$$

W should not, however, be made too small, because losses and large stray fields will result.

Also presented in Bosma's paper is an equation for the circulator bandwidth:

$$\frac{f_2 - f_1}{f_0} = 2.90 \frac{\kappa}{\mu} \rho \qquad (5.15)$$

where ρ is the maximum voltage reflection coefficient in the band, related to the circulator VSWR by Equation (2.4). The bandwidth is proportional to the ratio κ/μ, which also describes the amount of splitting between the resonant frequencies due to the two counter-rotating modes. When the splitting is greater, there is not so much variation in bandwidth-reducing quantities. The implication of Equation (5.15) to achieve large bandwidths is to operate the circulator very near to the ferrimagnetic resonance, which leads to high insertion loss.

Bosma defines the circulator input impedance as

$$Z = Z_0 - jZ_0 \frac{1.38}{\dfrac{\kappa}{\mu}} \frac{f - f_0}{f_0} \qquad (5.16)$$

where f is the frequency where the impedance is to be evaluated, and Z_0 is the stripline characteristic impedance. We can see that this impedance behaves like a parallel resonant circuit: inductive for $f < f_0$ and capacitive for $f > f_0$. The paper states that the bandwidth of the circulator can be increased by adding series-resonant circuits at each of the three ports.

Fay and Comstock [4] describe the stripline circulator junction as being equivalent to a transmission cavity with one port isolated. They state that two counter-rotating modes are present in the magnetized ferrite disks and the standing wave patterns are like the ones in Figures 5.2 and 5.4. The two rotating modes are said to have opposite reactive components of impedance, so that the net impedance at the operating frequency is real. The directions of circulation are opposite for below- and above-resonance operating points.

This paper also presents electromagnetic field equations for the biased ferrite. Equations (5.6) and (5.7) are asserted and the solutions of the field equations are given as

$$J_0(kR) - \frac{J_1(kR)}{kR} (1 + \kappa/\mu) = 0 \qquad (5.17)$$

$$J_0(kR) - \frac{J_1(kR)}{kR} (1 - \kappa/\mu) = 0 \qquad (5.18)$$

for the lowest order, $N = 1$ modes. The signs in front of the κ/μ terms correspond to the positive and negative rotating modes.

Fay and Comstock discuss the variation in mode splitting (κ/μ) with frequency relative to the ferrimagnetic resonance frequency (ω_0/ω).

Mode splitting refers to the amount of frequency splitting (difference in frequency) between the two resonance frequencies of the counter-rotating modes in the ferrite. The counter-rotating modes or waves have the same resonant frequencies when no dc magnetic field is applied to the ferrite, but when a magnetic bias is applied, the propagation constants of the two modes are no longer equal and the resonance frequencies become different. Insofar as the circulator operates between the two resonant frequencies, the circulator bandwidth is related to the amount of mode splitting. As described previously in this section, the operation of the junction circulator is based on the two counter-rotating wave components or modes.

Splitting is greatest at frequencies above ferrimagnetic resonance (below-resonance magnetic operating point). This indicates that greater bandwidths can be obtained using below-resonance circulators. Fay and Comstock also observe that the amount of splitting varies rapidly with saturation magnetization in the below-resonance region, hence the choice of $4\pi M_s$ is more critical for below-resonance circulators.

The Fay and Comstock paper also contains Equation (5.8) as a solution of the field equations for the degenerate case. The wave admittance of the circulator at the center frequency of operation is

$$G_r = \frac{Y_{\text{eff}}|\kappa/\mu|}{\sin \psi} \tag{5.19}$$

where Y_{eff} is the intrinsic wave admittance:

$$Y_{\text{eff}} = \sqrt{\frac{\varepsilon\varepsilon_0}{\mu_0 \mu_{\text{eff}}}} \tag{5.20}$$

The paper describes the derivation of the loaded Q of the circulator junction from the energy stored in the ferrites and the amount of power coupled to the striplines. The result is

$$Q_L = 1.48 \frac{\omega R^2 \varepsilon\varepsilon_0}{G_r d} \tag{5.21}$$

where d in the denominator of this equation is the thickness of the ferrite disks. Fay and Comstock go on to develop expressions for the unloaded Q of the ferrite resonators that could be used to approximate the insertion loss of the circulator junction. For most below-resonance circulators, the unloaded Q is given by

$$Q_u = \frac{1}{\dfrac{\gamma^2 4\pi M_s \, \Delta H}{2\omega^2} + \tan \delta} \tag{5.22}$$

where $\tan \delta$ is the dielectric loss tangent of the ferrite material. The first term in the denominator of Equation (5.22) represents the magnetic losses. The exact expression for the unloaded Q due to magnetic losses is $\mu'_{\text{eff}}/\mu''_{\text{eff}}$, which is difficult to evaluate without the approximations used.

The insertion loss of the circulator junction is approximately given by

$$\text{Insertion loss (dB)} = 10 \log_{10} (1 - Q_L/Q_u) \tag{5.23}$$

This paper asserts that the rotating modes should each have 30 degree phase angles at the circulator center frequency. Because of this, the splitting of the modes is related to the loaded Q by

$$\frac{\omega_+ - \omega_-}{\omega} = \frac{\tan 30°}{Q_L} = \frac{0.577}{Q_L} \tag{5.24}$$

Fay and Comstock describe the electrical equivalent circuit of the circulator input impedance as a parallel-resonant one. For narrow bandwidths, the circulator VSWR is related to the junction admittance and its phase angle as follows:

$$\text{VSWR} = \frac{Y_r^2}{G_r^2} = \sec^2 \theta \tag{5.25}$$

The circulator synthesis procedure suggested in this paper starts with the specified VSWR for the circulator, from which we find the admittance phase angle from Equation (5.25). Next, we calculate the required Q_L from

$$Q_L = \frac{\tan \theta}{\dfrac{f_2 - f_1}{f_0}} \tag{5.26}$$

Once Q_L is known, we can find the required amount of counter-rotating mode frequency splitting using Equation (5.24). Because the splitting is approximately proportional to κ/μ, we relate κ/μ to Q_L with a constant:

$$\frac{\kappa}{\mu} = \frac{0.71}{Q_L} \tag{5.27}$$

The required κ/μ can sometimes be realized in both the above- and below-resonance operating regions if the frequency is in the zone where the two regions overlap. The next step in the synthesis procedure would be to select the operating region and ferrite material based on the criteria presented in Chapters 2 and 4. Once the saturation magnetization of the ferrite is known, the disk radius is calculated using Equations (5.6), (5.7), and (5.8). A simplification of Equation (5.6) applies if the circulator is operated in the below-resonance region and the ferrite is just-saturated (internal magnetic field = 0):

$$\mu_{\text{eff}} = 1 - \kappa^2 = 1 - \frac{\gamma^2 M_0^2}{\omega^2} \qquad (5.28)$$

Fay and Comstock suggest the use of a quarter-wavelength impedance transformer to match the circulator junction to the system characteristic impedance (typically 50 Ω). From the known bandwidth and phase angle of the junction admittance (Equation (5.25)), a characteristic impedance for the quarter-wavelength transformer can be determined, either by graphical means with a Smith chart or numerically. After the transformer impedance is known, the circulator junction conductance is found by transformation of the system characteristic admittance through the quarter-wavelength transformer. The last step in the circulator synthesis procedure is to calculate the ferrite disk thickness using Equation (5.21).

The Fay and Comstock paper goes on to briefly discuss H-plane and E-plane waveguide junction circulators. The H-plane circulator is constructed from a symmetrical waveguide junction in the H-plane, with ferrite material placed in the center as shown in Figure 5.6. The ferrite is generally excited by the transverse RF H-field in the waveguide, because the rectangular waveguide usually operates in the TE_{10} mode and the ferrite is small in comparison to the width of the waveguide. The amount of coupling to the ferrite can be limited by placing irises at each of the three waveguides where they enter the center section of the circulator. The center section then becomes the circulator junction and the air in this area must be included in the analysis. If the irises are not used, the boundaries of the center cavity are not clearly defined, which leads to some of the difficulties involved in the analysis of waveguide junction circulators. Reduced-height (low impedance) waveguide is often used to construct these circulators.

E-plane circulators contain ferrites that are excited by the longitudinal RF H-fields. Usually, no ferrite is present in the center of the junction because it would not be active magnetically. This device is very different from the stripline junction circulator in that the striplines have no longitudinal RF H-fields.

Helszajn [5] describes an approximation of Q_L as a function of κ/μ in his paper, "Quarter-Wave Coupled Junction Circulators Using Weakly Magnetized Disk Resonators." He claims that the approximation is valid for values of κ/μ up

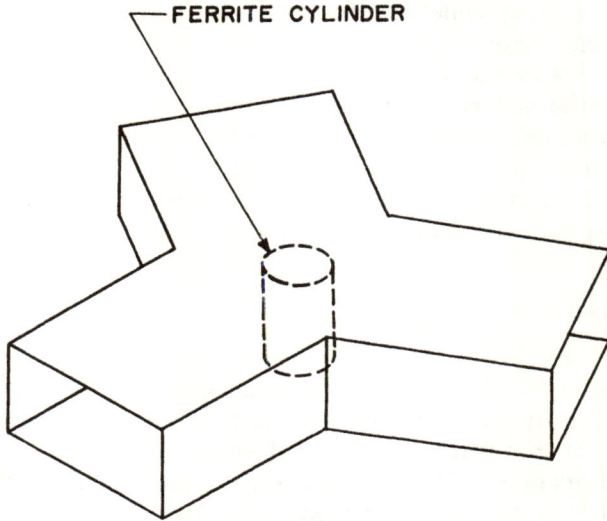

Fig. 5.6 *H*-plane waveguide junction circulator construction.

to 0.3, which corresponds to a minimum achievable Q_L of about 2. The fact that the theory applies to weakly magnetized ferrites implies that it is only valid for below-resonance designs.

He states that Q_L is nearly independent of ψ, considering the conditions imposed. The approximation for Q_L is derived from the equations presented by Bosma, Fay, and Comstock, and includes a second-order polynomial correction factor to make the classic values of Q_L match other theoretical Q_L values calculated by other means. The approximation is

$$Q_L = 0.689 \, \frac{\mu}{\kappa} + \left(0.0463 - 2.6318 \, \frac{\kappa}{\mu} + 3.5513 \, \frac{\kappa^2}{\mu^2} \right) \tag{5.29}$$

Helszajn presents equations for the parameters of the complex gyrator equivalent circuit of the circulator:

$$G_r = \frac{Y_0(r - \sin^2 \theta)}{r \cos^2 \theta} \tag{5.30}$$

$$B' = \frac{\pi Y_0}{4} \frac{(r - \sin^2 \theta)^{1/2}}{r \cos \theta} (r - 1)\tan^2 \theta \tag{5.31}$$

$$Q_L = \frac{\pi}{4} \frac{(r-1)\sin\theta\tan\theta}{(r-\sin^2\theta)^{1/2}} \tag{5.32}$$

where Y_0 is the system characteristic admittance (usually 0.02 S), r is the circulator input VSWR, and θ is defined as follows:

$$\theta = \cos^{-1}\left\{0.707\cos\left[\frac{\pi}{2}\left(1 + \frac{f_2 - f_1}{2f_0}\right)\right]\right\} \tag{5.33}$$

The paper presents the relations of Q_L, B, and G_r to the circulator junction parameters:

$$G_r = \frac{0.0192\sqrt{\varepsilon}}{\sqrt{\mu_{\text{eff}}}\, kR \sin\psi \ln\left(\frac{W+t+2d}{W+t}\right)} \frac{\kappa}{\mu} \tag{5.34}$$

$$B = \frac{0.0111\sqrt{\varepsilon}}{\sqrt{\mu_{\text{eff}}}\sin\psi \ln\left(\frac{W+t+2d}{W+t}\right)} \frac{J_1'(kR)}{J_1(kR)} \tag{5.35}$$

$$B' = \frac{0.0111\sqrt{\varepsilon}}{\sqrt{\mu_{\text{eff}}}\sin\psi \ln\left(\frac{W+t+2d}{W+t}\right)} \frac{(kR)^2 - 1}{2kR} \tag{5.36}$$

$$Q_L = 0.689 \frac{\mu}{\kappa} \tag{5.37}$$

where t is the thickness of the stripline center conductor. At the circulator center frequency, $B = 0$ because $J_1'(kR) = 0$. B' is the susceptance slope parameter. Equation (5.37) is the classic approximation for Q_L, which is replaced by Equation (5.29) in Helszajn's paper.

Insofar as this paper is about weakly magnetized ferrites, we again apply the approximation of Equation (5.28). A more accurate expression for B' can be obtained by deriving the product of Equations (5.34) and (5.29).

The paper offers equations for the design of a quarter-wavelength impedance transformer to match the circulator junction impedance to the system characteristic impedance. One of these is

$$Z_t = (rG_r Y_0)^{-1/2} \tag{5.38}$$

G_r in this equation is calculated from the specifications for the circulator (Equation (5.30)). After G_r, B', and Q_L are calculated from the circulator specifications using Equations (5.30) to (5.32), the junction parameters are adjusted until the results from Equations (5.29), (5.34), and (5.36) match the previously calculated values.

"Design Data for Radial-Waveguide Circulators Using Partial-Height Ferrite Resonators," a paper by Helszajn and Tan [6], presents design information for H-plane waveguide junction circulators. The three geometries considered in the paper are shown in Figure 5.7. Figure 5.7(a) depicts a half-wavelength long ferrite cylinder open-circuited at both ends, 5.7(b) shows two quarter-wavelength ferrite cylinders coupled in the middle of the waveguide junction, and a single quarter-wavelength ferrite resonator is shown in Figure 5.7(c). The analysis of all three geometries is very similar. The performance of all three is also similar, but (c) is the simplest and uses the least amount of ferrite. The geometry shown in Figure 5.7(c) also has better parameter flexibility than the other two.

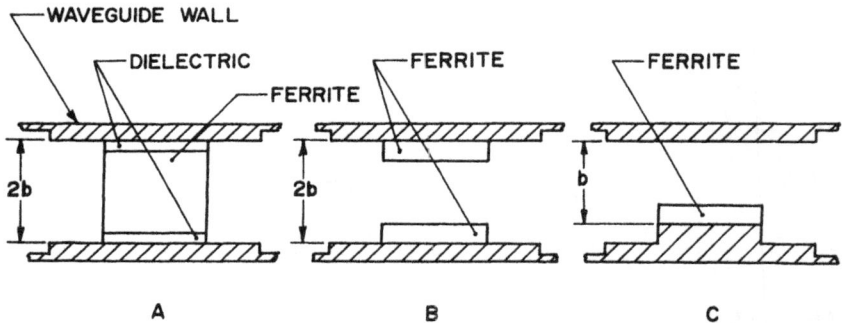

Fig. 5.7 Waveguide junction circulator geometries (edge views).

Helszajn and Tan state the circulator boundary conditions in terms of scattering matrix coefficients. They assert that $kR = 1.84$ for the degenerate case of the counter-rotating modes. The configurations described support axially resonating modes associated with the HE_{11} dielectric waveguide mode. The junction resonances can also be quantified using the theory of open dielectric resonators. The solution recommended in the paper is that of the TM_{111} resonator. The equation to be used for selection of resonator R and d dimensions is

$$k_0 R = \frac{1}{\sqrt{\varepsilon}} \left[\left(\frac{\pi R}{2d} \right)^2 + (1.84)^2 \right]^{1/2} \tag{5.39}$$

where $k_0 = 2\pi/\lambda_0$. The value of d is the ferrite disk thickness in Figures 5.7(b) and (c), and half the ferrite cylinder length in Figure 5.7(a). There are many possible values of d and R that will satisfy Equation (5.39). For large R/d, the resonance frequency depends mainly on d.

Where an image plane (an imaginary plane between the two coupled ferrites in Figure 5.7(b) or a waveguide wall is in close proximity to a ferrite resonator, the resonant frequency will be perturbed. Helszajn and Tan suggest a correction factor for the resonator length (d):

$$\Delta d = \frac{1}{\beta} \tan^{-1} \left\{ \frac{(k^2 - k_0^2)^{1/2}\varepsilon}{\beta} \tanh[(k^2 - k_0^2)^{1/2}s] \right\} - d \tag{5.40}$$

where β is a wave number that can be approximated by $\pi/2d$. The spacing between resonators or between resonator and image plane is represented by s.

The concept of the characteristic plane is discussed in this paper. The characteristic plane has several useful properties. First, it determines the point in the H plane where the impedance transformer should be connected. The admittances at each port of the circulator junction relative to the characteristic planes are related in a specific manner. Another property of the characteristic plane is that if there is a short circuit placed at this plane of one port, a sympathetic short circuit will also occur at the characteristic planes of the other ports. If a short circuit is placed at the output characteristic plane of a circulator, the input signal will be entirely reflected and the third port theoretically nulled.

The locations of the characteristic planes are specified by R_0, the radial distance from the center of the circulator junction to the plane. The value of k_0R_0 is approximately 3.45 when $k_0R = 1$, and k_0R_0 does not vary greatly with k_0R, but climbs slightly when k_0R goes above or below 1; k_0R_0 does not usually exceed 4.

Helszajn and Tan indicate that a suitable value for the physical dimensional ratio d/b is between 0.7 and 0.85.

In the paper, "Planar Triangular Resonators with Magnetic Walls," Helszajn and James [7] present field patterns for triangular ferrite resonators and magnetic field equations obtained by duality from TE mode equations.

The $TM_{1,0,-1}$ mode is dominant for a triangular resonator, and this is the mode for the counter-rotating circulator modes. The resonance frequency for this mode can be calculated from

$$kA = 3.63 \tag{5.41}$$

where A is the altitude or height of the triangle. The value of A needs to be corrected, say Helszajn and James, to compensate for fringing because the edges of the triangle are not ideal magnetic walls:

$$A_{eff} = A + \frac{d}{4} \tag{5.42}$$

Again, d is the ferrite thickness. Equation (5.42) applies when $A/d > 4$.

The paper presents some experimental data for two circulators: one built with a disk resonator, and one built with a triangular resonator. The data show that the insertion loss of the triangular junction is about 17% lower than the disk junction loss.

The paper by Davies [8] is a circulator analysis that describes matching cylindrical modes in the ferrite to associated modes in the waveguide junction outside the post, then in turn matching these modes to the rectangular waveguide modes. This results in a large number of equations and unknowns, but because some waveguide modes can be ignored, approximations emerge.

Perfect circulation is predicted for certain values of ferrite radius, κ, and μ. The basic concepts presented in this paper are the same as those already summarized, but the approach of mode matching is a little different.

Auld [9] presents a scattering-matrix method of synthesizing waveguide junction circulators. He asserts that once a particular symmetry for the circulator has been selected, the nature of the adjustments required to obtain the specified performance are suggested by the symmetry, and the range of adjustment must be experimentally determined.

Auld proposes that adjustments to the circulator junction should take the form of bending and denting the waveguide walls and placing any type of material in the junction. He later states that only the ferrite radius and magnetic field need to be adjusted.

According to Auld, the ferrite radius should be determined experimentally. This is done without magnetic bias, adjusting the ferrite radius until the VSWR is 2:1. The radius is then readjusted after magnetic bias is applied to achieve circulation.

He states that the adjustment of the circulator (determination of ferrite radius) is simplified if a metal pin is added axially through the center of the ferrite. Auld suggests that the ferrite should be rotated to cancel asymmetries.

Simon [10] describes the experimental development of a series of octave-bandwidth circulators covering 0.6–8.0 GHz. These are all below-resonance stripline devices.

He writes of the development process as primarily an impedance-matching problem. His first step was to study the impedance characteristic of the basic circulator junction, varying ferrite and structural parameters to arrive at a combination that was easy to match. For a broad bandwidth match, the reactive component of the impedance must be as low as possible.

The striplines, including those between the ferrite disks, were chosen to have a characteristic impedance of 50 Ω in air. The ground-plane spacings for the

1–2 GHz and 2–4 GHz circulators were 0.500 and 0.265 inches, respectively. The stripline center conductors had geometries as depicted in Figure 5.8.

The ferrite disk diameters suggested by Simon for the octave-bandwidth circulators are presented in Figure 5.9 in graphical form, as he presents them.

His second step in the development process was to develop matching structures to achieve the desired octave-bandwidth performance. The impedance characteristics from the first step were adjusted to be approximately resistances, lower in value than 50 Ω. Simon found that quarter-wavelength transformers (one at each port) were all that was required to achieve 20 dB isolation, 0.5 dB insertion loss, and 1.25:1 VSWR over the octave bandwidth for most of the circulator junctions. The lowest frequency circulator also required the use of a quarter-wavelength open-circuited stub, to cancel the reactive component of the junction.

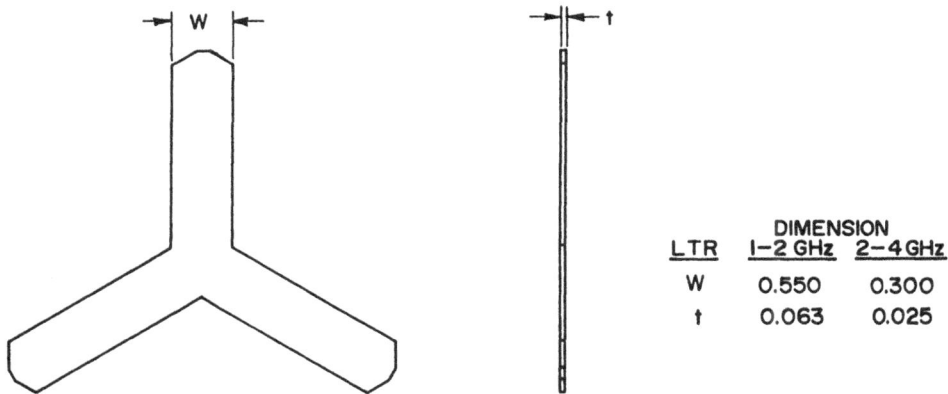

LTR	DIMENSION	
---	1–2 GHz	2–4 GHz
W	0.550	0.300
t	0.063	0.025

Fig. 5.8 Stripline center conductor geometries used by Simon [10].

Green and Sandy [11] characterized the microwave permeability of some partially magnetized ferrites and present this data along with empirical equations that closely match the measured data in their paper.

The permeability measurements were made using ferrite rods in cylindrical TM_{110} cavities. Although junction circulators do not normally utilize partially magnetized ferrites, the equations presented are useful for the design of other ferrite devices.

The first equation is from Rado:

$$\kappa' = \frac{\gamma 4\pi M}{\omega} \tag{5.43}$$

where $4\pi M$ is the ferrite magnetization, not to be confused with the saturation

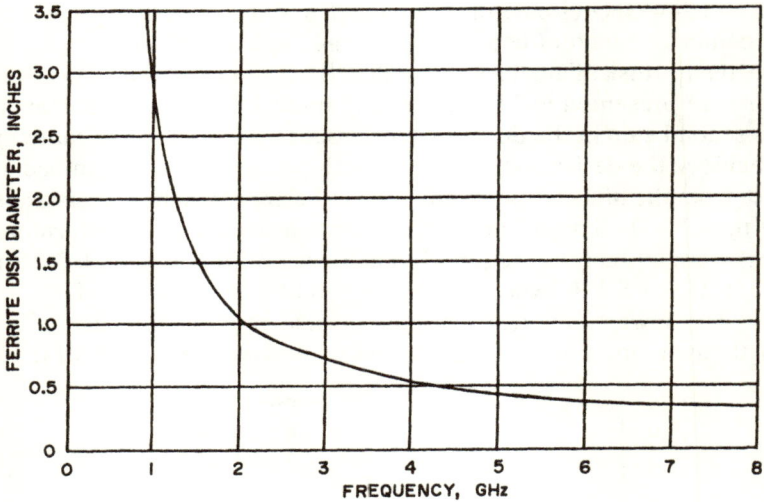

Fig. 5.9 Suggested ferrite disk diameters for octave-bandwidth circulators from Simon [10].

magnetization. The permeability for the completely demagnetized state, by Schloemann, is

$$\mu_0' = \frac{2}{3}\left[1 - \left(\frac{\gamma 4\pi M_s}{\omega}\right)^2\right]^{1/2} + \frac{1}{3} \tag{5.44}$$

Finally, Green and Sandy present two equations for which no physical basis is claimed. Because the ferrite is not saturated, the permeability in the direction the dc magnetic field is applied (z-axis) is not 1:

$$\mu_z' = \mu_0'\left(1 - \frac{4\pi M}{4\pi M_s}\right)^{5/2} \tag{5.45}$$

$$\mu' = \mu_0' + (1 - \mu_0')\left(\frac{4\pi M}{4\pi M_s}\right)^{3/2} \tag{5.46}$$

Some of the variable symbols in the papers summarized here do not match the symbols anywhere in this book; we want to keep variable symbols internally consistent.

Having completed the descriptions of some of the more important papers regarding the synthesis of circulators, we now proceed to develop design procedures for several classes of junction circulators.

First, we will consider the above-resonance stripline junction. Bosma's paper [3] deals with the above-resonance device, so we utilize some of the equations he presented. If the ferrites are magnetically biased far above resonance, it is true that $\omega_0^2 \gg \omega^2$ because ω_0 is proportional to H by virtue of Equation (1.7). Insofar as α is proportional to the ferrimagnetic resonance losses (Equation (1.9)), we can also write $\omega_0^2 \gg \alpha^2$ due to the fact that we have chosen our magnetic operating point far above resonance. Using the preceding assumptions and Equation (1.30), ignoring the loss component, which should be insignificant (Equation (1.31)), and substituting (1.7) γH_{dc} for ω_0, we arrive at Equation (5.10):

$$\mu = \mu_{eff} = \frac{H_{dc} + M_0}{H_{dc}} \qquad (5.10)$$

Because κ is much smaller than μ for practical above-resonance operating points, we reduce Equation (5.6) (also a factor in Equation (1.37)) to $\mu_{eff} = \mu^2/\mu = \mu$.

We know that the ferrite disk is a resonator, so we can find its radius if we know the wavelength for which we want it to be resonant, the ferrite permeability (μ_{eff}), and the ferrite dielectric constant. From the circulator specifications we know the wavelength, and from the specifications for the ferrite material chosen we know the dielectric constant. We must calculate the effective permeability from H_{dc} and M_0.

Insofar as the ferrite material will be saturated by the dc magnetic field, the ferrite magnetization (M_0) will equal the ferrite saturation magnetization ($4\pi M_s$). Having already selected ferrite material, we know the saturation magnetization. We now select a value of magnetic field, H_{dc}. Bosma suggests that the field should be at least four times greater than the field required for resonance, so we adopt a rule of thumb with the help of Equation (1.7):

$$H_{dc} = \frac{4\omega}{\gamma} = 1.4 f_0 \text{ (MHz)}, \quad \text{Oe} \qquad (5.47)$$

Two other considerations influence the choice of H_{dc}: the desired bandwidth and maximum allowable insertion loss. If H_{dc} is set too low, resonance losses will occur at the upper-band edge of the circulator response. Figure 5.10 shows insertion loss *versus* H_{dc} for several values of resonance line width. A saturation magnetization of 1800 G was used to calculate these curves, but this is not a critical parameter. For an above-resonance circulator, the resonance losses are highest at the upper-band edge, so this frequency should be used to evaluate the loss.

We calculate the losses using Equations (1.30) to (1.33), (5.6), and (5.23).

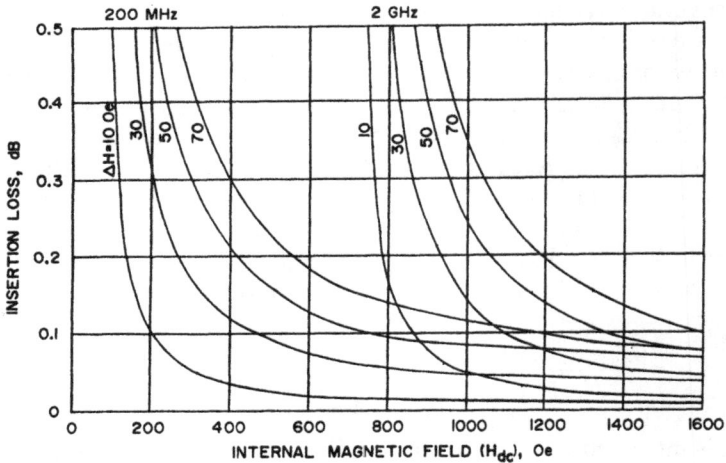

Fig. 5.10 Insertion loss *versus* internal magnetic field. $4\pi M_s = 1800$ G.

The unloaded Q is $\mu'_{\text{eff}}/\mu''_{\text{eff}}$ and Q_L is approximated by Equation (5.27).

The bandwidth without any impedance transformers or matching circuits is related to the amount of splitting between the resonant frequencies of the counter-rotating modes by Equation (5.15):

$$\frac{f_2 - f_1}{f_0} = 2.90 \frac{\kappa}{\mu} \rho \tag{5.15}$$

From Equations (5.10) and (1.27) we can derive an approximation for κ/μ based on the previously stated assumptions:

$$\frac{\kappa}{\mu} = \frac{M_0 \omega}{\gamma H_{\text{dc}}(H_{\text{dc}} + M_0)} \tag{5.48}$$

We see from this expression that decreasing H_{dc} will increase bandwidth (but also increase insertion loss), as will increasing M_0.

Once we have established H_{dc}, we proceed to calculate the ferrite disk radius using Equations (5.9) and (5.10). Next, we find the required loaded Q (Q_L) and resonator conductance (G_r) from the specified circulator VSWR and bandwidth with Equations (5.30), (5.32), and (5.33). Fay and Comstock also presented equations to calculate Q_L from bandwidth and VSWR, but the equations due to Helszajn are useful for a broader range. The ferrite disk radius can be read from Figure 5.11, where it is plotted against μ_{eff} for several values of dielectric constant.

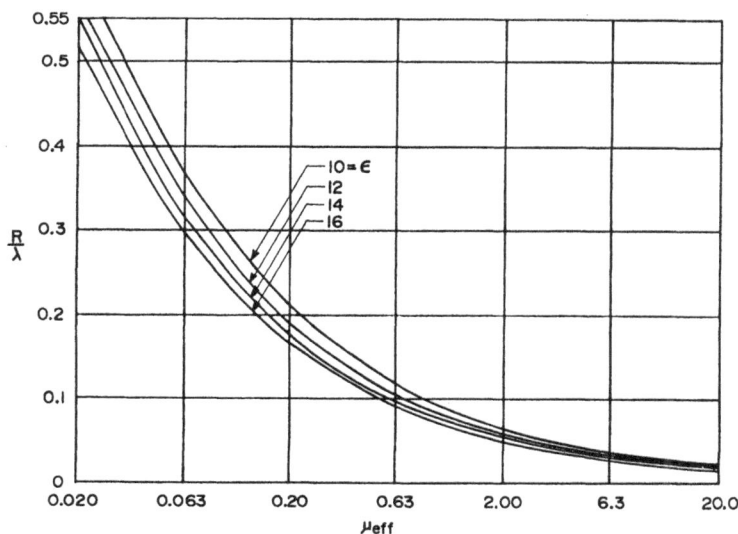

Fig. 5.11 Disk radius *versus* μ_{eff}.

Curves for Q_L and G_r *versus* bandwidth with VSWR as a parameter are shown in Figures 5.12 and 5.13, respectively. The bandwidth in Equations (5.30), (5.32), and (5.33), and in the figures is the bandwidth that can be achieved using quarter-wavelength impedance-matching transformers.

From Q_L and G_r, we compute the ferrite disk thickness using Equation (5.21). We must avoid two pitfalls here. One is that the ferrite may be so thin that it cannot be manufactured. The solution to this problem is to increase the bandwidth, which will result in a thicker ferrite disk. The advantage of a narrow bandwidth is low loss in the ferrite, but this may be negated by high circuit losses because the high junction conductance requires impedance-matching circuitry having inherently higher losses.

The other pitfall is that the disk may be so thick, and hence the ground plane spacing so large, that higher-order modes can propagate. The cut-off wavelength of the first higher-order mode is approximately given [12] by

$$\lambda_c = \sqrt{\varepsilon}\,(2W + 3.172d) \tag{5.49}$$

Another approximation is to see that $d \ll \lambda/4$ at the operating frequency of the circulator.

The stripline width (W) at the circulator junction has an effect on the amount of coupling to the ferrite. A wider strip increases the coupling because it couples

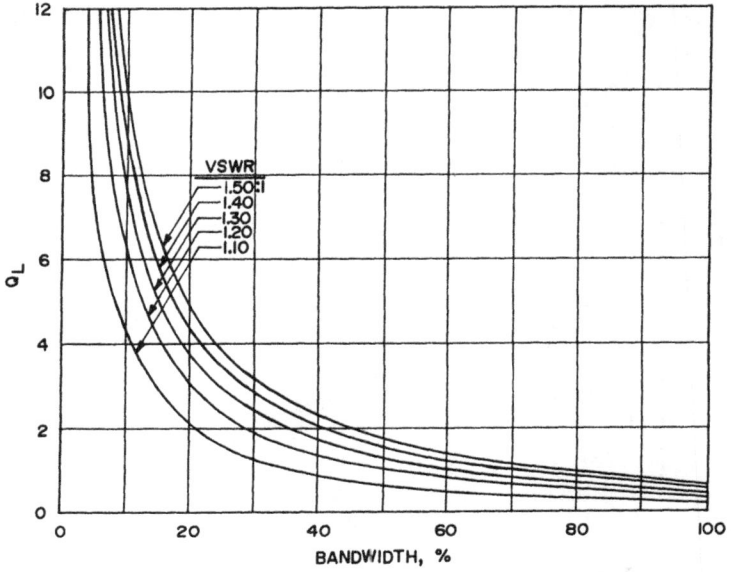

Fig. 5.12 Ferrite disk resonator loaded Q *versus* bandwidth.

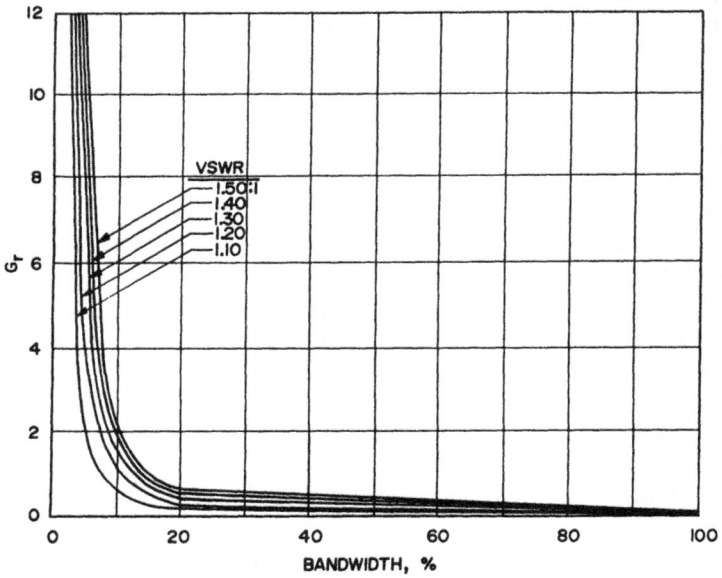

Fig. 5.13 Ferrite disk junction conductance *versus* bandwidth.

into the two counter-rotating modes over a wider range of frequencies. The angular position of the nodes and antinodes of the standing-wave pattern in the ferrite disk changes with frequency, so a wider strip accommodates a wider range of frequencies.

If the width is chosen too wide, the RF magnetic field in the ferrite is substantially reduced, which is undesirable. A narrow strip width may result in high circuit losses and large stray fields.

As a starting point, we introduce another rule of thumb: set W equal to the width that gives a stripline characteristic impedance of 50 Ω, assuming air dielectric. This rule is presented without proof, as its basis is largely empirical. Equations (5.13) and (5.14) should hold true for the chosen strip width.

The stripline characteristic impedance can be determined [13] using one of two sets of equations:

$$Z_0 = \frac{60}{\sqrt{\varepsilon}} \ln \frac{4b}{\frac{W}{2}\left[1 + \frac{t}{\pi W}\left(1 + \ln \frac{4\pi W}{t} + 0.51\pi \frac{t^2}{W^2}\right)\right]\pi} \tag{5.50}$$

valid for $W/b < 0.35$ and $t/b < 0.25$; t is the stripline thickness and b is the ground-plane spacing ($b = 2d$ for small t). For $W/b > 0.35$,

$$Z_0 = \frac{94.15}{\left(\frac{W/b}{1 - t/b} + \frac{C_f'}{0.0885\varepsilon}\right)\sqrt{\varepsilon}} \tag{5.51}$$

where

$$C_f' = \frac{0.0885\varepsilon}{\pi}\left[\frac{2}{1 - t/b} \ln\left(\frac{1}{1 - t/b} + 1\right)\right.$$

$$\left. - \left(\frac{1}{1 - t/b} - 1\right) \ln\left(\frac{1}{(1 - t/b)^2} - 1\right)\right] \tag{5.52}$$

Many reference books present curves for stripline Z_0, and microwave circuit design software packages can be used to compute stripline parameters.

For the below-resonance operating region, we use magnetic fields lower than the fields for above-resonance circulators. We make the simplifying assumption that the internal magnetic field in the ferrite is 0. This is the case when the ferrite is just magnetized. Dropping the loss terms, we derive Equation (5.28) from Equations (5.6), (1.30), and (1.32):

$$\mu_{\text{eff}} = 1 - \frac{\gamma^2 M_0^2}{\omega^2} \tag{5.28}$$

Note that $\mu = 1$.

We find the ferrite disk radius in the same manner as we do for above-resonance circulators, using Equation (5.9) or Figure 5.11 and Equation (5.28). The effective permeability, hence the disk radius, is only a function of the ferrite material selected and the operating frequency of the circulator. The magnetic field does not have a strong influence on the loss and bandwidth as it does with the above-resonance junction, but the ferrite saturation magnetization is more critical.

We can compare the calculated disk radius with the disk radii that Simon empirically found to be optimum. These radii are presented in Figure 5.9. The disk radii found by either method should be similar, but the radii due to Simon will, without a doubt, lead to working octave-bandwidth designs.

Next, we compute Q_L and G_r the same way we do for the above-resonance circulator, using Equations (5.30), (5.32), and (5.33) or Figures 5.12 and 5.13. We approximate the required κ/μ using Equation (5.27).

If $\kappa/\mu < 0.3$, we can use Helszajn's design expressions. We find a more exact value of κ/μ from Equation (5.29). If $\kappa/\mu > 0.3$, or we do not wish to calculate a more accurate κ/μ, we simply use the previously approximated value from Equation (5.27).

We present two equations that relate Q_L and κ/μ because each has advantages. Clearly, both are approximations, as are most circulator design expressions. Equation (5.27) is the simpler of the two, hence easier to use. It also provides reasonably accurate results over a fairly wide range of values. Helszajn's expression, Equation (5.29), includes a second-order polynomial correction factor that improves accuracy, but only for $\kappa/\mu < 0.3$. It is more difficult to use because of the polynomial factor, particularly if we are solving for κ/μ.

The ferrite thickness should be arbitrarily set to about $\lambda/16$ (wavelength in air), which is the approximate dimension that Simon used. The ferrite thickness is not critical as other adjustments later on such as stripline width, and center conductor geometry will compensate for inaccuracies in disk thickness.

For the case of $\kappa/\mu < 0.3$, we calculate the stripline width from Equation (5.34). Otherwise, we again set W equal to the width that gives a stripline characteristic impedance, with air dielectric, of 50 Ω.

There are several other design considerations that apply to both above-resonance and below-resonance circulators. One of these is the impedance transformer. If the specified circulator bandwidth dictates that one or more quarter-wavelength impedance transformers are needed, we must design these transformers. The circulator bandwidth without any transformers or matching circuits is given by Equation (5.15). Most octave-bandwidth below-resonance circulators contain at least one quarter-wavelength transformer at each port, and often two are used.

We compute the characteristic impedance of a single transformer using Equation (5.38). If two transformers are used, the impedances are given by

$$Z_{t1} = G_r^{-3/4} Y_0^{-1/4} \qquad (5.53)$$

$$Z_{t2} = G_r^{-1/4} Y_0^{-3/4} \qquad (5.54)$$

where Z_{t1} is the characteristic impedance of the transformer nearest the junction.

To minimize discontinuities and transformer size, it is a good idea to set the dielectric constant of the dielectric surrounding the transformer center conductor equal to or close to the ferrite dielectric constant. Transformers may be folded to minimize the amount of space they occupy.

Lumped-element impedance-matching techniques can be used in low-frequency circulators. The important thing to remember is that the equivalent circuit of the circulator junction is typically a parallel-resonant circuit.

The ferrite disk junctions can easily be converted to their equilateral triangle equivalents by using the following conversion, derived from Equations (5.41) and (5.42):

$$A = 1.97R - \frac{d}{4} \qquad (5.55)$$

valid for $A/d > 4$. The thickness of the triangle is the same as the disk thickness.

A wide variety of center conductor geometries are utilized in commercial circulators, some of which are depicted in Figure 5.14. For the novice circulator designer, a simple disk between $3R/2$ and $2R$ diameter will work well with a ferrite disk, and for the ferrite triangle, a hexagon that just fits inside the triangle is a good starting point.

Power-handling capacity specifications will affect not only the choice of ferrite material for below-resonance circulators but also ground-plane spacing for both above- and below-resonance devices. At high peak powers, we are concerned with voltage breakdown if the center conductor is too close to the ground plane. The theoretical breakdown power [12] of a stripline with rounded edges and $t/b > 0.05$ is at least $600 \text{ kW}/b^2$, where b is in inches and $Z_0 = 50 \ \Omega$. Generally, the highest characteristic impedance present in a stripline circulator is $50 \ \Omega$, because the junction impedance is usually lower. The RF voltage is higher for higher characteristic impedances, so the worst case from a voltage-breakdown point of view is when $Z_0 = 50 \ \Omega$.

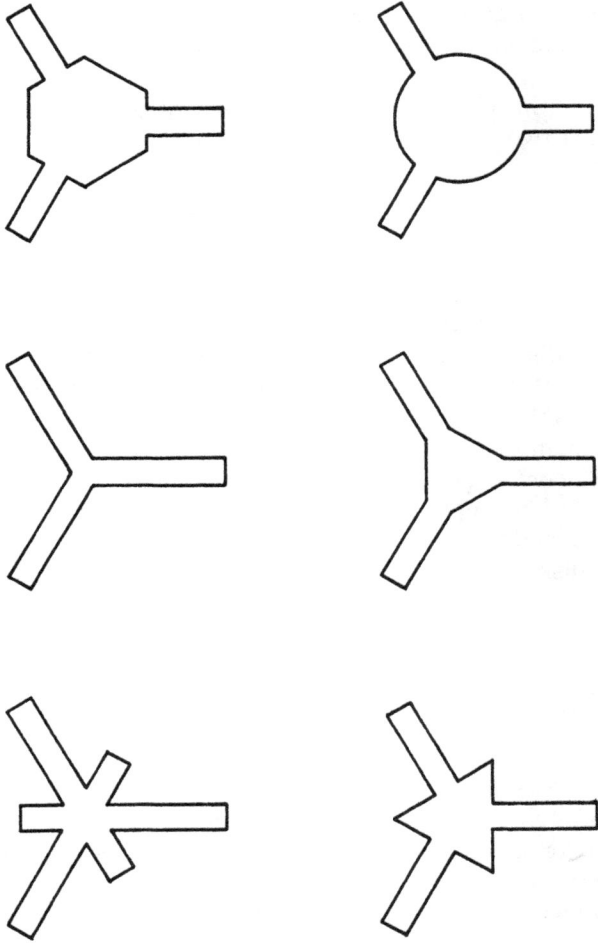

Fig. 5.14 Center conductor geometries.

Another method of calculating the stripline breakdown power is to use Equation (2.10), substituting 70 kV/d for V_{dw}, where d is the ferrite or dielectric thickness.

If the ground-plane spacing cannot be increased, the alternatives are to completely fill the space between the center conductor and ground planes with a dielectric material that has a high dielectric strength (such as silicone) or to pressurize the circulator.

The concern when high average powers are present is that the ferrite, dielectric material, center conductor, or solder joints will overheat. We analyze the striplines between the circulator junction and the connectors by first determining the insertion loss of these lines, the amount of power dissipated by the line per unit length, and finally the temperature rise of the center conductor and dielectric near it, considering that the heat is conducted through the dielectric to the ground plane. This analysis is beyond the scope of this book, but some of the thermal considerations are discussed in Chapter 7.

The critical component of the circulator to be designed for high average power is the ferrite junction. The amount of power dissipated in the ferrites must be minimized. This means we want low insertion loss. To keep the surface of the ferrite nearest the center conductor cool, we need to use thin ferrites. This leads to a smaller ground-plane spacing and higher losses in the striplines. Also, an extremely thin ferrite disk (or triangle) may produce higher loss in the ferrite, hence more power dissipated in the ferrite. In addition, other parameters, such as bandwidth and VSWR, may degrade if the ferrite is made too thin. Thus, we have a trade-off between electrical performance and thermal performance.

After an optimum ferrite thickness has been chosen, we have still another method of reducing the surface temperature of the ferrite. This method is to reduce the power density in the ferrite by increasing the disk diameter or triangle altitude, while not increasing the insertion loss proportionally. The only way we can increase the surface area of the ferrite without changing the operating frequency of the circulator is to decouple the center conductor from the disk. We can also operate the ferrite resonators in higher-order modes, the next mode occurring with disk diameters about 1.6 times larger.

We explain the concept of decoupling by thinking of the ferrite disks (or triangles) and the immediately surrounding volume as a cavity resonator. If this abstract cavity contains no ferrite (only air) it will have a particular resonant frequency for a given mode. If ferrite is added to the cavity, the resonant frequency will generally decrease due to the permittivity and permeability of the ferrite, which are greater than the values for air. The ratio of ferrite volume to air volume in the cavity is described by a quantity called the filling factor. The more ferrite we have in a cavity of given size, the higher the filling factor. A higher filling factor implies a lower resonant frequency.

A larger cavity will also have a lower resonant frequency than a smaller cavity. Our goal is to increase the ferrite volume without changing the operating frequency of the circulator. To do this, we must necessarily increase the cavity volume because the "cavity" of a typical stripline junction circulator is fully filled with ferrite. If we increase the cavity volume without changing the filling factor, the circulator frequency will decrease. We therefore reduce the filling factor to maintain the same circulator operating frequency.

We implement the reduction in filling factor by substituting dielectric material having a lower dielectric constant than ferrite for some of the ferrite. It is desirable to have the ferrites in contact with the ground planes for good thermal conduction (to keep them cool), so we place the dielectric between the ferrite and the center conductor.

The disadvantage of decoupling the center conductor from the disks is that the loaded Q of the junction will increase, which implies a decrease in bandwidth. In general, decoupling a resonant circuit from source and load will increase circuit Q.

Decoupling is accomplished by placing dielectric material between the ferrite and the center conductor. A flexible material such as Teflon works well; a good starting thickness is about 0.05 inches. The decoupling scheme is illustrated in Figure 5.15.

The transitions from coaxial line to stripline need attention. Figure 5.16 shows the details of a typical transition. Both coax and stripline operate in the *TEM* mode, so what is required of the transition is to make a smooth change from the symmetrical coax mode to the distorted stripline mode. Ideally, we would like the stripline ground-plane spacing to be equal to the outer diameter of the coaxial dielectric. When the stripline spacing is larger than this diameter, the transition is fairly good without extensive compensations. When the ground-plane spacing is smaller, we need to add an intermediate section to step down the size of the center conductor. This section can take the form of a round hole bored into the side of the circulator where the connector attaches. The center conductor inside this bore can be either round or flat, like the stripline. Transition design for the circulator can be borrowed from other types of stripline devices. It is difficult to analyze transitions accurately, so the design must be accomplished by cut-and-try methods.

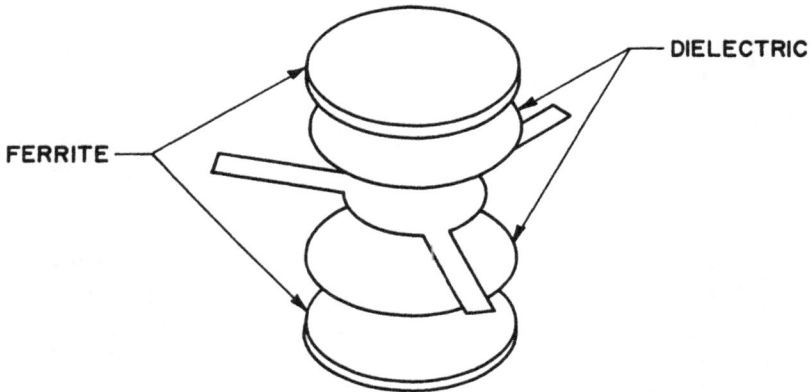

Fig. 5.15 Ferrite decoupling scheme.

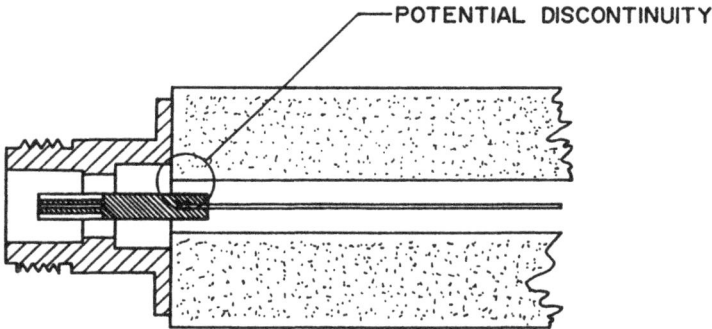

Fig. 5.16 Coaxial-stripline transition.

An important concept in circulator design is that of frequency scaling. Because the disk radius or triangle altitude is proportional to wavelength, as are magnetic field and saturation magnetization, designs can be scaled. By this, we mean that if we have a working design with center frequency f, we can generate another design with center frequency f' that has very similar (if not identical) electrical performance. If we denote the new design parameters by R', M_0', and H_{dc}', and the old design parameters by R, M_0, and H_{dc}, we can write

$$R' = \frac{f}{f'} R \tag{5.56}$$

$$M_0' = \frac{f'}{f} M_0 \tag{5.57}$$

$$H_{dc}' = \frac{f'}{f} H_{dc} \tag{5.58}$$

M_0 need not be changed for an above-resonance design, and H_{dc} for a below-resonance device need only be changed to just saturate the ferrite. This scaling technique is only valid for small changes in frequency (less than 1.5:1), and we do need to change other parameters along with the ones in Equations (5.56) to (5.58).

We will briefly discuss two more classes of junction circulators that are not as widely used as the stripline junction circulators: the microstrip circulator and the H-plane waveguide junction circulator.

The waveguide circulator design procedure we will present applies to the geometries shown in Figure 5.7. The first step is to calculate k_0, the wave number for the center frequency of the circulator. This parameter is calculated using

$$k_0 = \frac{2\pi}{\lambda_0} \qquad (5.59)$$

Next, we select a ferrite thickness (d) between $0.7b$ and $0.85b$. The dimension b in Figure 5.7(c) and $2b$ in Figures 5.7(b) and 5.7(a) can be taken as the waveguide height, or the junction may be constructed in a reduced-height center section. Reducing the waveguide height slightly increases the junction conductance, hence decreases the bandwidth and the insertion loss.

We compute the ferrite disk radius with the aid of Equation (5.39). The last step is to find R_0, the radius where the impedance transformers (if used) connect to the junction area. R_0 should be about 3.5 times R.

Smaller values of d can be selected at the expense of electrical performance. The bandwidth of the waveguide junction circulator is more narrow than that of the stripline circulator because of the inherently narrow bandwidth of waveguide.

Several techniques have been used to improve the performance of waveguide junction circulators, including the addition of dielectric sleeves around the ferrite disks and metal pins through the center of the ferrite.

Most microstrip circulators exhibit no clear differentiation between the junction portion and the impedance-matching structure. The radius of the junction is usually interpreted [14] as the magnet radius. The magnet is normally in contact with one face of the ferrite substrate, opposite the microstrip circuitry. The radius of the junction area (and of the magnet) is calculated in the usual manner using Equation (5.9), except that the constant 1.84 is reduced to 1.65 because there is no clearly defined edge of the junction.

The thickness of the substrate is usually chosen to match the substrate thickness of the circuit with which the circulator is intended to be used. A center conductor geometry utilizing a disk with radius about $0.8R$ will work well, and microstrip line impedances should be kept as low as possible to achieve reasonable line widths.

Nearly all microstrip circulators operate below resonance. This allows the use of low magnetic fields.

The microstrip circuitry is normally deposited on the ferrite substrate using thin-film techniques. Sometimes, a dielectric ring is placed around a ferrite disk to form the microstrip circulator. This construction is more complicated and has no particular advantage.

5.2 LUMPED-CONSTANT CIRCULATORS

We learned in Section 5.1 that the size of the ferrite disk in a junction circulator is proportional to wavelength. This means that the size of the ferrites becomes prohibitively large at low frequencies. The ferrite size in a lumped-element circulator does not necessarily increase with wavelength.

The lumped-element circulator consists [15] of a ferrite disk with three coils wound on it so that the RF magnetic fields are oriented at 120 degrees with respect to each other. Figure 2.9 shows a ferrite disk with the center conductor (coil) wrapped around it. Other usable geometries are illustrated in Figure 5.17.

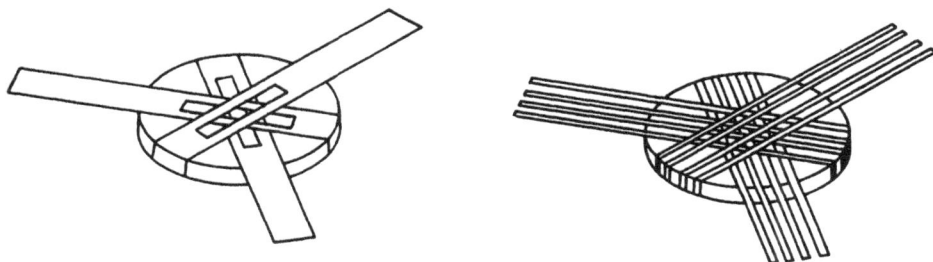

Fig. 5.17 Lumped-element circulator geometries.

The length of the coils is much less than a wavelength at the circulator operating frequency, so they are essentially inductances. The junction resonances are formed by connecting capacitances at the three ports. The capacitances can be connected either in the shunt or series configuration, as illustrated in Figure 5.18. The center node of the three nonreciprocally mutual coupled inductances can be either connected directly to ground or grounded through a series-resonant circuit. The series-resonant circuit, suggested by Konishi and Hoshino [16], improves the bandwidth of the circulator by canceling reactive components of the junction impedance at frequencies removed from the center frequency.

In the following discussion, we will consider the directly grounded junction, resonated with series capacitors [17]. The analysis of the shunt-capacitor case would be very similar.

The RF magnetic field in the ferrite disks is shown in Figure 5.19. We let the inductance of each coil, assuming no ferrite is present (inductance in air), equal L_0. If we think of the propagation in the ferrite as being parallel to the applied dc magnetic field, knowing that $L = \mu L_0$, we derive from the factor μ in Equation (1.40):

$$L_+ = L_0(\mu + \kappa) \tag{5.60}$$

$$L_- = L_0(\mu - \kappa) \tag{5.61}$$

where plus and minus signs denote the two counter-rotating modes. The mutual

SHUNT   SERIES

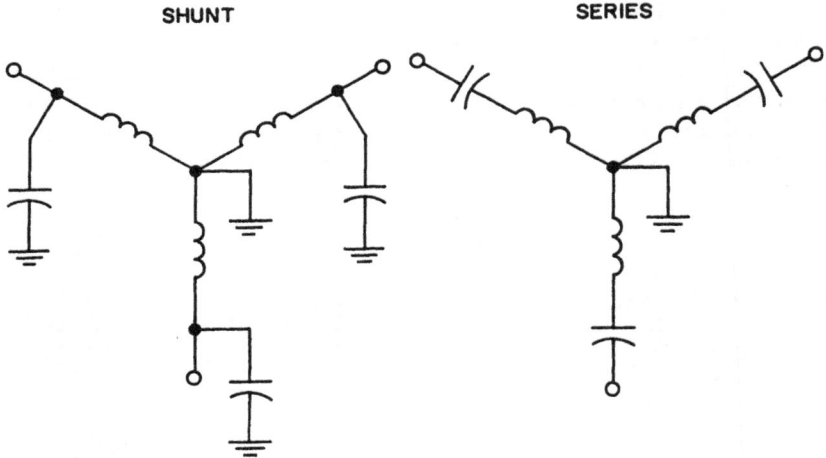

Fig. 5.18 Lumped-element circulator capacitor connections.

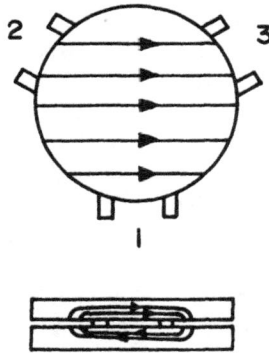

Fig. 5.19 RF magnetic field in lumped-element circulator ferrites. Port 1 is excited.

inductance of any pair of coils depends on the direction of signal propagation. We could also analyze the circulator junction using the transverse propagation case and suitable changes in the equations presented here, with little difference in the end result.

For ideal circulation, Dunn and Roberts [17] derive

$$\frac{\omega}{2}(L_+ - L_-) = \frac{Z_0}{1.73} \tag{5.62}$$

$$\frac{\omega}{2}(L_+ + L_-) = \frac{1}{\omega C} \tag{5.63}$$

where C is the capacitance of the series capacitors. We see from these equations that the inductance is an important parameter. The inductance can be achieved with a many-turn coil wrapped around a small ferrite disk or a one-turn coil wrapped around a large ferrite disk.

Neglecting losses, we derive expressions for μ' and κ' from Equations (1.30) and (1.32):

$$\mu' = 1 + \frac{\gamma M_0 \omega_0}{\omega_0^2 - \omega^2} \tag{5.64}$$

$$\kappa' = \frac{\gamma M_0 \omega}{\omega_0^2 - \omega^2} \tag{5.65}$$

By substituting Equations (5.60) and (5.61) into (5.62) and (5.63), we derive

$$1.73 \, \omega \, L_0 \kappa = Z_0 \tag{5.66}$$

$$\omega^2 L_0 \mu = \frac{1}{C} \tag{5.67}$$

The ferrite disks in a lumped-element circulator are biased above resonance. The techniques described in Section 5.1 for determining the magnetic operating point and bandwidth apply for the lumped-element circulator as well.

To design a lumped-element circulator, we first select an appropriate ferrite material based on the criteria presented in Chapter 4. Next, we choose a value of internal magnetic field, H_{dc}, referring to Equation (5.47), Figure 5.10, and Equation (5.15). κ/μ can be approximated using Equation (5.48) or Equations (5.64) and (5.65).

After we have fixed H_{dc} we can find L_0 from Equations (5.66) and (5.65). We also compute the series capacitor value (C) from (5.67) and (5.64).

The ferrite disk dimensions are somewhat arbitrary. The disk diameter must be much less than a wavelength, or the center conductor coils will not appear as pure inductances. A good rule of thumb is to set the disk diameter equal to $\lambda/16$ or smaller. A larger disk has two advantages: higher average power-handling capacity due to reduced power density in the ferrite, and a center conductor with a larger cross section for a given inductance. The disk thickness is even less important. A thin disk is advantageous both from a cost point of view and a thermal consideration. If the disk chosen is too thin, however, the center conductor cross section becomes vanishingly small, negating any power-handling advantages of the thin disk. A good starting point for disk thickness is one-tenth the disk diameter.

The width of the strips wrapped around the ferrite are selected to provide the previously calculated inductance in air (L_0). One strip could be used for each port, but usually two or more strips are connected in parallel. This is done to distribute more evenly the RF magnetic field in the ferrite. Paralleling also reduces the total inductance at each port.

The inductance of a strip in air can be calculated from the stripline impedance formulas (5.50) to (5.52) if we know that [18]

$$L = 0.08467\, Z_0, \quad \text{nH/in} \tag{5.68}$$

If we opt to use two strips at each port, the inductance of each strip must be twice L_0. We use $2L_0$ to find the required Z_0 in Equation (5.68), then adjust the strip width to achieve this Z_0.

The strips from adjacent ports are insulated from each other where they cross with an insulating tape such as Teflon (PTFE) tape. The insulation thickness should be kept to a minimum so that all the strips are close to the ferrites. The crossover points should also have low capacitance. The simultaneous requirements of thin spacing and low capacitance may seem to be conflicting, but they are not. There are two dimensional variables that apply to a parallel-plate capacitor model. One is the spacing between the capacitor plates, and the other is the area of the plates. We can change the amount of capacitance even if we choose to fix one of the dimensional variables—the spacing. To reduce the capacitance at the crossover points, we can make the strips more narrow in the vicinity of the cross. This may be necessary if wide strips are to be used. An analysis of the effects of this interstrip capacitance on circulator performance would be quite involved and, in nearly all cases, not justified.

The peak power-handling capacity of lumped-element circulators is limited because of the necessarily thin insulation on the strips.

In practice, lumped-element circulators are not perfectly symmetrical because the strips from one port are not against the ferrites but between the strips from the other two ports. This problem can be eliminated by weaving the center conductor so that all the strips have an equal amount of contact with both ferrites.

If sufficient bandwidth cannot be obtained with just the circulator junction, we add impedance matching (compensation) sections. Because of the low frequencies involved, these networks would be too large if they were constructed with sections of transmission line. Therefore, they are constructed from discrete capacitors and inductors. The circuits usually take the form of bandpass structures, having parallel- and series resonant sections. It is usually best to evaluate the junction alone before adding the compensation networks.

5.3 DIFFERENTIAL PHASE SHIFT CIRCULATORS

The operation of the waveguide differential phase shift circulator is described in Section 2.4. There are several methods of implementing the required differential phase shifts. The first method, shown in Figure 5.20, uses identical magic tees or quadrature couplers at each end of the two waveguide sections. One waveguide section has a 180 degree nonreciprocal (differential) phase shifter, and the other section is dielectrically loaded to produce a reciprocal (not differential) phase shift equal to the phase shift, in one direction, of the 180 degree section.

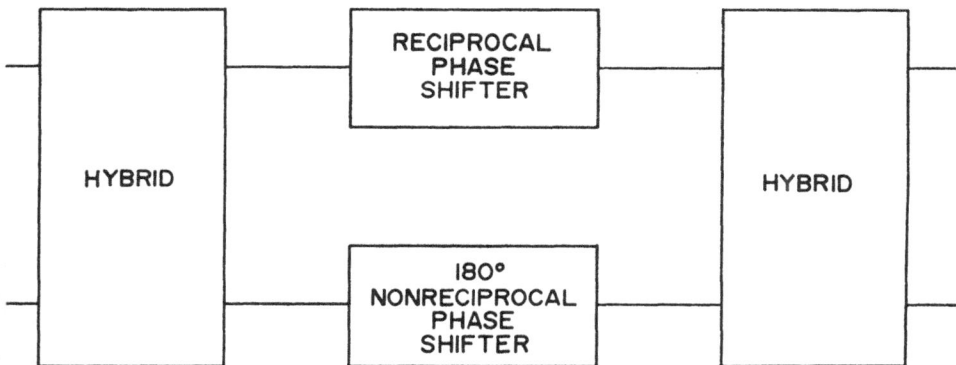

Fig. 5.20 Differential phase shift circulator implementation.

The second method employs 90 degree nonreciprocal phasers in both waveguide sections, between identical tees or couplers. One of the waveguide sections also contains a 90 degree reciprocal phase shifter. This reciprocal phaser could simply be a dielectric slab placed inside the guide. This method of implementation is illustrated in Figure 5.21.

Figure 5.22 shows that the third implementation method is to use 90 degree nonreciprocal phasers in both waveguide sections, but use different types of couplers at the ends. If a magic tee is used at one end of the waveguides, a quadrature coupler is used at the other end.

Design data for couplers and tees are readily available, and an entire book could be written just about these components, so we will not discuss their design here.

The basic geometry of the differential phase shifter is shown in Figure 5.23. From perturbation theory, Clarricoats [19] derives expressions for propagation constants of ferrite-loaded waveguides. These expressions contain nonreciprocal terms; that is, the wave traveling through the waveguide suffers different phase

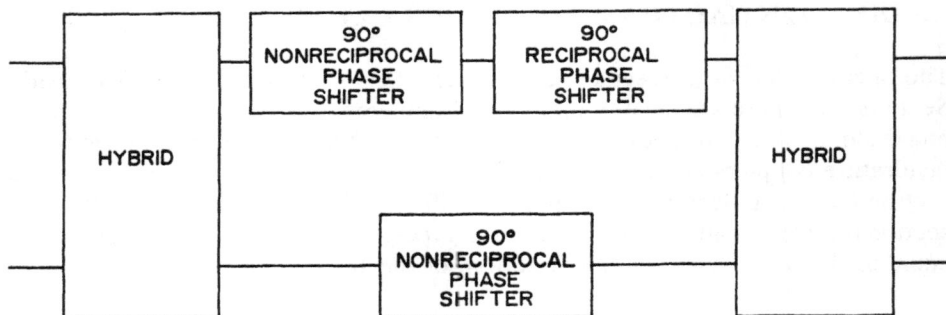

Fig. 5.21 Differential phase shift circulator implementation.

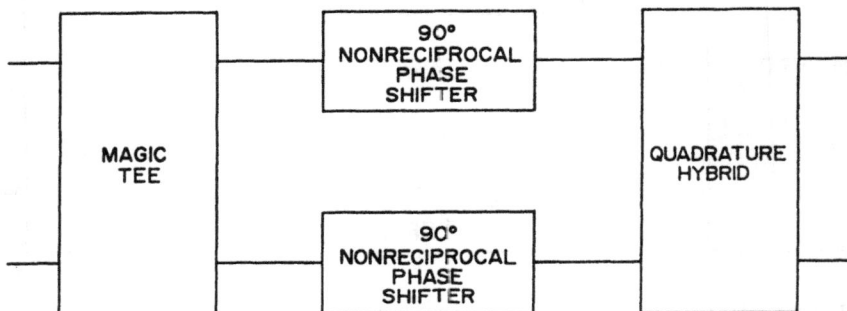

Fig. 5.22 Differential phase shift circulator implementation.

shifts depending on the direction of propagation. We have already discussed the reasons for this nonreciprocal behavior in Chapter 1. The only additional information needed is that the ferrite material is located in the waveguide such that the fields in the ferrite are circularly polarized.

There are natural regions of circular polarization in rectangular waveguides that operate in the TE_{10} mode. When ferrite slabs are placed in these regions, waves with circular polarization are excited in the ferrite. The sense of polarization (left or right hand) depends on the direction of wave propagation in the waveguide. Waves traveling in the forward direction in the waveguide excite circularly polarized waves in the ferrite with one sense of polarization, and waves traveling in the reverse direction excite waves in the ferrite with the opposite sense of polarization.

Because the propagation constants for the two different senses of circular polarization are different as described in Section 1.4, there are different phase

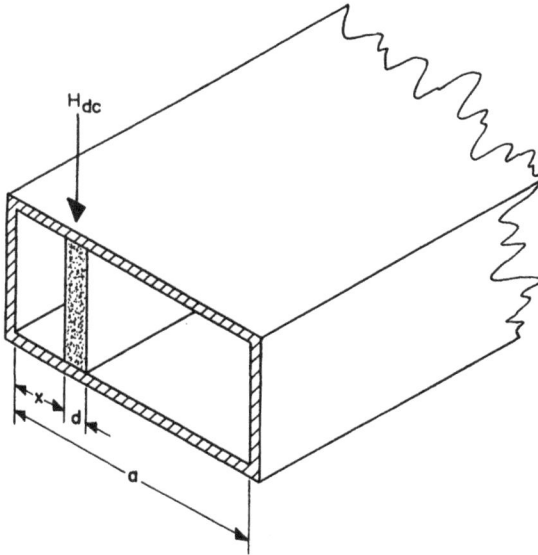

Fig. 5.23 Nonreciprocal waveguide phase shifter.

shifts associated with the transmission of the two types of waves in the ferrite-loaded waveguide.

The natural regions of circular polarization in rectangular waveguide occur, as illustrated in Figure 5.24. As a wave travels down the guide, the loops formed by the magnetic field lines also travel. If we consider the direction and magnitude of the magnetic field at a particular position and instant of time, we have the field vector shown in Figure 5.24. As time progresses, the field vector rotates due to the change in the magnetic field. At a certain position in the waveguide, the magnitude of the magnetic field vector does not change; only the direction changes, and we have circular polarization. Circular polarization occurs at a point in the guide about one quarter of the way across. The sense of polarization (direction of magnetic field vector rotation) is opposite for waves traveling in opposite directions in the waveguide.

The differential phase shift per unit length is given by

$$\beta_+ - \beta_- = \frac{2\pi d}{a^2} \frac{\kappa}{\mu} \sin \frac{2N\pi x}{a} \tag{5.69}$$

where d is the ferrite slab thickness, a is the waveguide width, and N is an integer (we let $N = 1$ for our purposes). The maximum differential phase shift is obtained

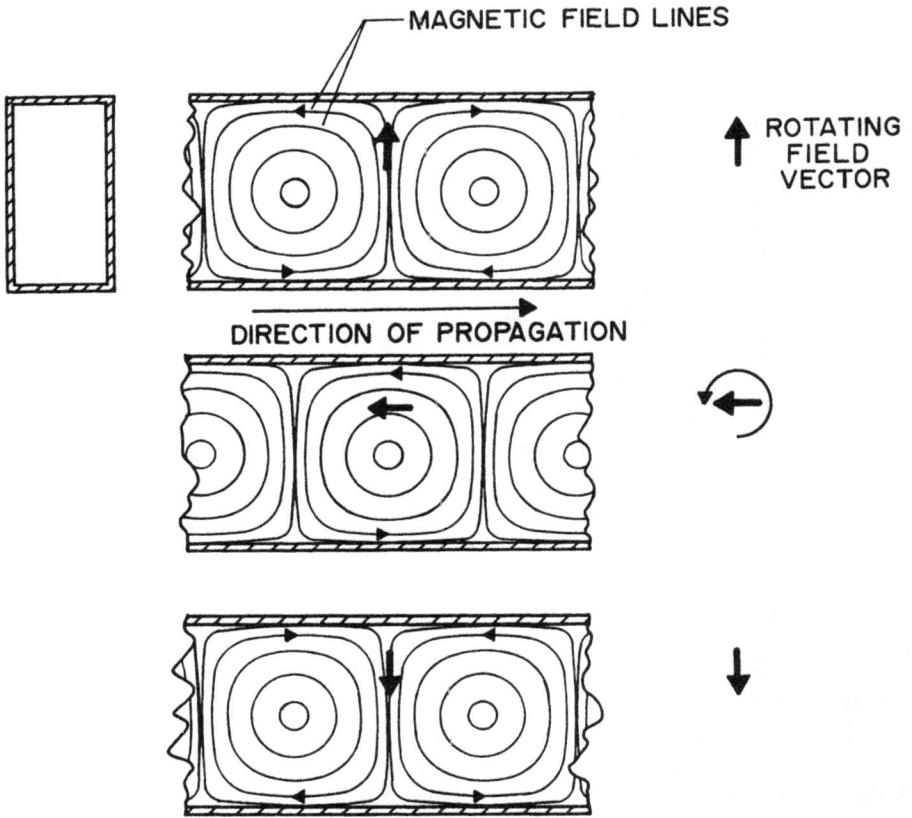

Fig. 5.24 Circular polarization in a rectangular waveguide.

when $x = a/4$:

$$\beta_+ - \beta_- = \frac{2\pi d}{a^2} \frac{\kappa}{\mu} \qquad (5.70)$$

The ferrite slab in a differential phase shifter is biased below resonance, typically in the region where $\mu = \kappa$. When $\mu = \kappa$, we find from Equations (1.30) and (1.32), assuming the ferrite is lossless, that

$$H_{dc} = \frac{\omega}{\gamma} - M_0 \qquad (5.71)$$

We see from Equation (5.70) that to achieve a high value of differential phase shift, we need $\mu < \kappa$. The factor κ/μ increases with magnetic field, but the insertion loss climbs as we approach ferrimagnetic resonance.

After selecting an appropriate ferrite material, we establish the dimensions of the slab. The thickness of the slab (d) should be greater than $a/10$ to maximize the differential phase shift, but less than $\lambda/10$ to avoid multimode propagation. The length of the slab is calculated using Equation (5.70), where the desired differential phase shift depends on the particular design but will usually be 90 degrees. The value of κ/μ can be determined by numerical analysis of the insertion loss as a function of H_{dc} as described in Section 5.1. The highest value of κ/μ is obtained with the highest tolerable magnetic field from an insertion-loss point of view. As an alternative to the computations, we can assume $\kappa/\mu = 1$, remembering that the differential phase shift will be higher than the value we calculate using this assumption.

The position x of the ferrite slab is somewhat critical, but should be near $a/4$. There is no nonreciprocal phase shift when the ferrite slab is placed against the waveguide wall or in the center of the guide. The position should be adjusted experimentally to obtain good differential phase shift and, more importantly, minimum variation in phase shift over the frequency band of interest.

The isolation the finished circulator will have as a function of variation in differential phase shift is approximated by

$$\text{Isolation} = -10 \log_{10} \left\{ \sin \left[\frac{\Delta(\beta_+ - \beta_-)}{2} \right] \right\} \tag{5.72}$$

To reduce reflections (VSWR) from the ferrite slab, the ends can be tapered. The length of the taper should be determined experimentally. The bandwidth of the phaser can be increased by dielectric loading techniques, which are also largely empirical in nature.

Improved thermal performance and, consequently, improved power handling can be achieved by using thin ferrite slabs mounted against the guide's broad walls instead of the slab oriented in the E plane, shown in Figure 5.23. This scheme is illustrated in Figure 5.25.

5.4 RESONANCE ISOLATORS

The resonance isolator is a two-port nonreciprocal attenuator that relies on the absorption of energy at ferrimagnetic resonance. The basic configuration of the resonance isolator is very similar to the differential phase shifter geometry shown in Figure 5.23.

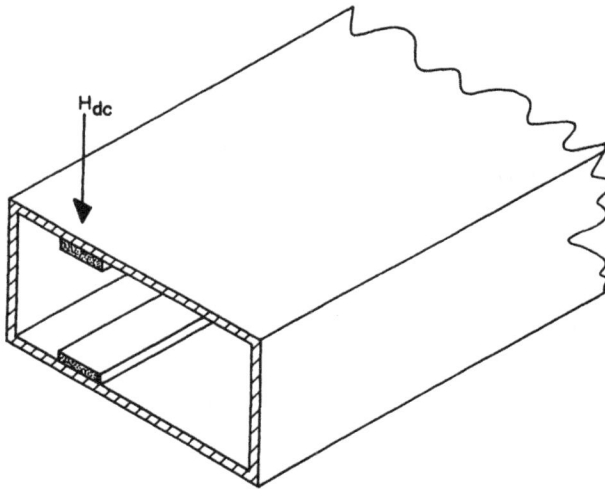

Fig. 5.25 Alternative ferrite slab location.

The position of the ferrite slab in the waveguide is such that resonance absorption occurs for only one direction of propagation, as described in Section 2.6. Forward and reverse waves traveling through the waveguide excite circularly polarized waves in the ferrite slab with opposite senses of polarization. The rotation of the reverse circularly polarized mode coincides with the direction of electron precession, so the energy in the reverse wave is coupled into the precession system, and we say it undergoes resonance absorption. Because the electrons can only precess in one direction for a given polarity of applied dc magnetic field, and the direction of rotation of the forward circularly polarized mode does not coincide with the direction of electron precession, the forward wave does not exhibit resonance absorption.

The mechanism of circular polarization in a rectangular waveguide is explained in Section 5.3. The ferrite slab in a resonance isolator is located in a region of circular polarization in the waveguide, so elliptically polarized waves can propagate in the ferrite.

An improvement in the front-to-back ratio of the isolator can be obtained by placing one to four ferrite slabs with their faces against the broad walls of the waveguide, as shown in Figure 5.25. This configuration also increases the average power-handling capacity because the slabs are in good thermal contact with the walls. A disadvantage of not filling the entire height of the guide with ferrite is that a much higher magnetic field is necessary to bias the ferrite slabs. This problem can be minimized, however, by using reduced-height waveguide.

Various methods have been suggested [19] to increase the bandwidth of resonance isolators. One is to use several ferrite slabs with different saturation magnetizations, one for each band of frequencies. Another method is to use nonuniform applied magnetic fields or different ferrite shapes.

Isolation and front-to-back ratio can be improved by employing dielectric loading techniques. In a waveguide isolator, the best location for the dielectric is against the ferrite face nearest the center of the guide, as shown in Figure 5.26.

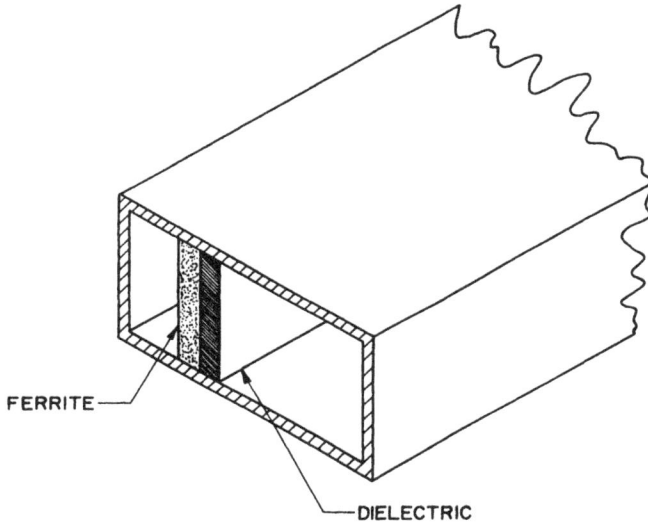

FERRITE

DIELECTRIC

Fig. 5.26 Dielectric loading.

The magnetic operating point of the resonance isolator is, of course, at resonance. Equation (1.7) describes the relation between resonance frequency and internal magnetic field.

The guidelines presented in Section 5.3 for ferrite thickness and positioning also apply to the resonance isolator, but the exact geometry must be determined experimentally.

We must use dielectric material in the construction of coaxial resonance isolators. The dielectric helps create a circularly polarized field at the dielectric-air interface, which would not be present without the dielectric. A cross-sectional view of a coaxial resonance isolator is shown in Figure 5.27.

The stripline junction isolator offers better electrical performance, smaller size, and lower cost (in many cases) than the coaxial resonance isolator. For these reasons, the junction isolator is much more popular. We will not present detailed design information for the coaxial resonance isolator because it is unlikely that

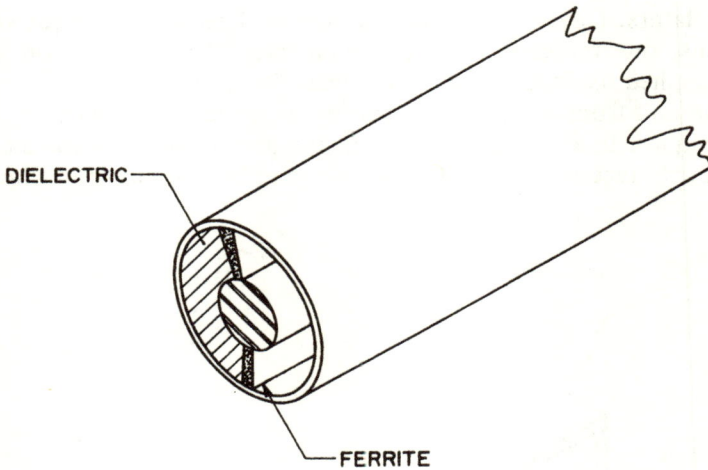

Fig. 5.27 Coaxial resonance isolator.

anyone designing modern equipment will need to design one of these devices. Several other types of circulators have been omitted because they are no longer built; one example is the Faraday rotation circulator.

5.5 DUMMY LOADS FOR ISOLATORS

Dummy loads, also called terminations and RF loads, are often integrated with circulators to form isolators. Therefore, we will discuss some basic design concepts for these components.

Loads are made for both coaxial and waveguide transmission media. Some typical coaxial loads are shown in Figure 5.28 and waveguide terminations in Figure 5.29.

The absorptive element in a coaxial load could be either a thin-film resistor or bulk absorbing material such as polyiron. Thin-film resistors are made with beryllium oxide substrates, which have very good thermal conductivity. As an example of the average power-handling capacity of these resistors, a rod $\frac{3}{8}$ inch diameter by $\frac{3}{4}$ inch long can dissipate up to 60 W continuously if properly cooled. Rod resistors are sold in a wide variety of sizes.

Polyiron, as we call it, is an iron-loaded epoxy-based material that absorbs RF energy. The material is available in different attenuation values; the attenuation is measured using a fully filled length of coaxial line. The attenuation is specified in dB per unit length at a specific frequency, usually 10 GHz. The attenuation is approximately proportional to frequency. The power-handling capacity of this material is typically about 40 W/in at room temperature.

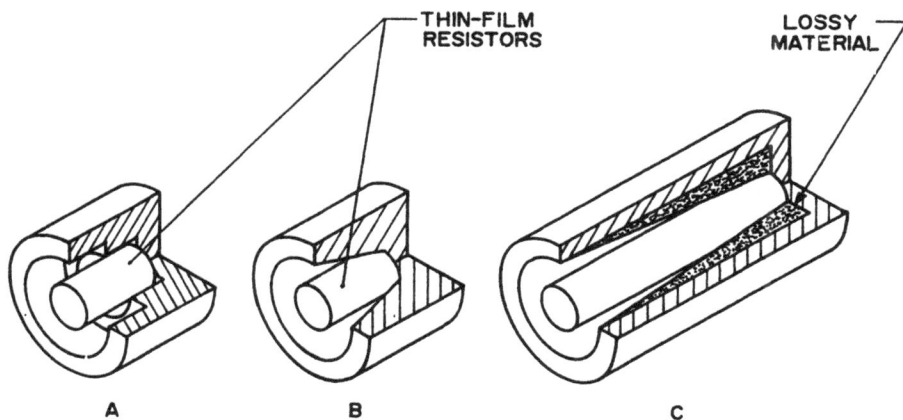

Fig. 5.28 Typical coaxial terminations.

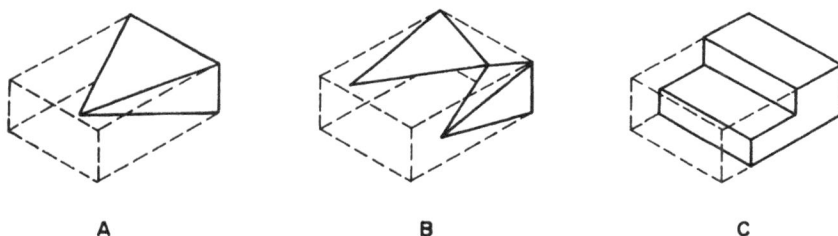

Fig. 5.29 Typical waveguide terminations.

The key parameters of an RF load are the VSWR, frequency range, and power-handling capacity. To achieve high average power handling, it is important to keep the absorptive element as cool as possible, because the RF energy is converted to heat. The rod resistors have a quite broad frequency range, extending from dc to the frequency where the skin depth is much less than the thickness of the resistive film. As we know, most of the current in the inner conductor of a coaxial line flows at its surface. The skin depth is the depth from the surface where the current decreases to $1/e$ of the current at the surface.

For low VSWR, the outer conductor around the rod resistor must be tapered. This tapering is done to maintain a characteristic impedance (between the rod and outer conductor) equal to the resistance of the rod resistor to ground at any point along its length. When the wavelength is much larger than the length of the resistor, a simple step taper like the one shown in Figure 5.28(a) will suffice. The small diameter forms a 25 Ω Z_0, and the larger diameter is 50 Ω. The best taper is logarithmic, as shown in Figure 5.28(b).

The coaxial loads built with absorptive material (polyiron) have more narrow bandwidths because of the frequency sensitivity of the material. For low VSWR (< 1.10:1), the taper illustrated in Figure 5.28(c) should be at least one wavelength long, and the fully filled portion of the line should be at least one outer-conductor diameter long. The taper serves not only to provide a good impedance match, but also to distribute the power more evenly in the absorber. This type of load is usually limited in bandwidth to about one octave. We choose the material so that it has sufficient loss at the lowest frequency in the operating band to achieve the required return loss. The upper frequency limit is then set by the nonuniform thermal distribution in the absorber.

The waveguide absorptive elements can be made from either polyiron or a high-temperature ceramic material. The ceramic (refractory) material has greater power-handling capacity than the epoxy material. Of the three tapers shown in Figure 5.29, the one shown at 5.29(b) has the highest power capacity because it has the greatest contact with the metallic waveguide walls, which can conduct away the heat. The step taper shown in Figure 5.29(c) has the advantage of being small in size and easy to manufacture, but does not offer good bandwidth because the transition to the material is not smooth. The shape in 5.29(a) is easy to remove and install in the guide because it only touches one wall.

Again, a taper length of one wavelength will provide a good match over the full waveguide bandwidth.

A wide variety of commercial RF loads are on the market, many of which are easily integrated with a circulator and are cost effective.

REFERENCES

1. Soohoo, R. F., *Microwave Magnetics* (New York: Harper and Row, 1985).
2. *Tech-Briefs* (Adamstown, MD: Trans-Tech, 1973), p. 16.
3. Bosma, H., "On Stripline Y-Circulators at UHF," *IEEE Transactions on Microwave Theory and Techniques,* January 1964, pp. 61–73.
4. Fay, C. E., and R. L. Comstock, "Operation of the Ferrite Junction Circulator," *IEEE Transactions on Microwave Theory and Techniques,* January 1965, pp. 1–13.
5. Helszajn, J., "Quarter-Wave Coupled Junction Circulators Using Weakly Magnetized Disk Resonators," *IEEE Transactions on Microwave Theory and Techniques,* May 1982, pp. 800–806.
6. Helszajn, J., and F. C. Tan, "Design Data for Radial-Waveguide Circulators Using Partial-Height Ferrite Resonators," *IEEE Transactions on Microwave Theory and Techniques,* March 1975, pp. 288–298.
7. Helszajn, J. and D. S. James, "Planar Triangular Resonators with Magnetic Walls," *IEEE Transactions on Microwave Theory and Techniques,* February 1978, pp. 95–100.
8. Davies, J. B., "An Analysis of the M-Port Symmetrical H-Plane Waveguide Junction with Central Ferrite Post," *IRE Transactions on Microwave Theory and Techniques,* 1962, p. 596.
9. Auld, B. A., "The Synthesis of Symmetrical Waveguide Circulators," *IRE Transactions on Microwave Theory and Techniques,* April 1959, pp. 238–246.

10. Simon, J. W., "Broadband Strip-Transmission Line Y-Junction Circulators," *IEEE Transactions on Microwave Theory and Techniques*, May 1965, pp. 335–345.
11. Green, J. J., and F. Sandy, "Microwave Characterization of Partially Magnetized Ferrites," *IEEE Transactions on Microwave Theory and Techniques*, June 1974, pp. 641–645.
12. Matthaei, G. L., L. Young, and E. M. T. Jones, *Microwave Filters, Impedance-Matching Networks, and Coupling Structures* (Dedham, MA: Artech House, 1980).
13. Howe, H., *Stripline Circuit Design* (Dedham, MA: Artech House, 1974).
14. Harrison, G. R., G. H. Robinson, B. R. Savage, and D. R. Taft, "Ferrimagnetic Parts for Microwave Integrated Circuits,'" *IEEE Transactions on Microwave Theory and Techniques*, July 1971, pp. 577–588.
15. Helszajn, J., *Passive and Active Microwave Circuits* (New York: John Wiley and Sons, 1978).
16. Konishi, Y., and N. Hoshino, "Design of a New Broad-Band Isolator," *IEEE Transactions on Microwave Theory and Techniques*, March 1971, pp. 260–269.
17. Dunn, V. E., and R. W. Roberts, "New Design Techniques for Miniature VHF Circulators," *International Microwave Symposium Digest*, 1965, pp. 147–151.
18. *Reference Data for Radio Engineers* (Indianapolis: Howard W. Sams, 1977).
19. Clarricoats, P. J. B., *Microwave Ferrites* (New York: John Wiley and Sons, 1961).

Chapter 6
Magnetic Design

6.1 MAGNET SIZING

All the circulators described in this book require dc magnetic fields. This magnetic bias causes the electrons in the ferrite material to precess as described in Section 1.3. This precession is essential to the operation of the circulator.

In some cases the magnetic bias is furnished by an electromagnet or by the retentivity of the ferrite itself. Usually a current pulse through a wire loop supplies the energy to set the amount and polarity of magnetic retention.

Ferrite retentivity is measured in gauss. The retentivity, or remanence, is the flux density in the ferrite when the external magnetic field is removed. Figure 6.1 is a typical hysteresis loop for a ferrite material. The retentivity is the amount of magnetic induction (flux density) where the loop crosses the B axis.

Switching circulators and phase shifters use retentive ferrites. The amount of energy we need to magnetize the ferrite can be determined from [1]

$$W = \int \left(N \frac{d\phi}{dt} + L \frac{di}{dt} + Ri \right) i dt \tag{6.1}$$

where ϕ is the magnetic flux, N is the number of turns of wire around the magnetizing core, i is the current flowing in the wire, L is the leakage inductance of the circuit, and R is the winding resistance. The circuit is shown schematically in Figure 6.2.

If the required magnetic flux is known, we can determine the amount of energy the current pulse must have using Equation (6.1). To change the energy level, we can change the wire loop (or coil) parameters. We calculate the required flux as described later in this section.

Where the dc magnetic field must be varied in an analog fashion, or retentive ferrites are not used, we use an electromagnet. The electromagnet is simply a coil

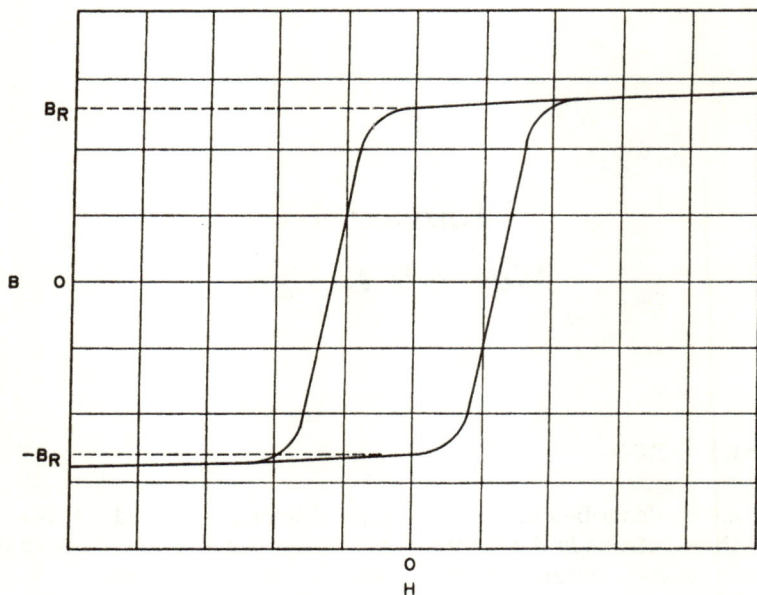

Fig. 6.1 Typical ferrite material hysteresis loop (B_R = retentivity).

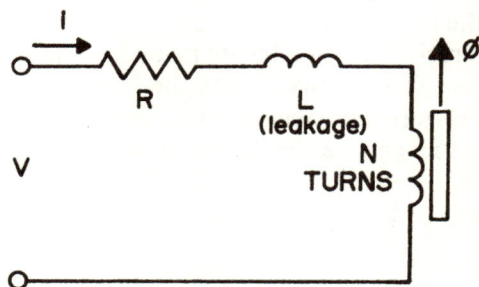

Fig. 6.2 Charging circuit for retentive ferrite devices.

of wire around an iron core. Ideally, the core should have a high permeability and a high saturation flux density. The magnetomotive force delivered by an electromagnet is equal to the product Ni (ampere turns). We will later discuss how to determine the required magnetic field from which the electromagnet coil can be designed. The procedure for designing the coil is much the same as that for inductor or transformer design. Insofar as this design information is readily available, we will not present it here.

In general, the circulator dc magnetic field is provided by permanent magnets. The choice of magnet material was described in Chapter 4; in this section we will concern ourselves with the sizing of the magnets.

A typical junction circulator magnetic circuit is shown in Figure 6.3. We introduce several components here that we have not previously described: shields, pole pieces, shunts, and returns. The purpose of the pole pieces is to make the magnetic flux density in the ferrites more uniform. Without pole pieces, the lines of flux are concentrated near the edges of the magnets, as shown in Figure 6.4. It is very important for the magnetic field applied to the ferrites to be uniform, or homogeneous. If the field is not homogeneous, portions of the ferrites can be only partly magnetized or, worse yet, not magnetized. This situation leads to high insertion loss and generally poor circulator performance. Pole pieces homogenize the magnetic field by conducting magnetic flux from strongly magnetic areas of the magnet to areas that would otherwise be weakly magnetized. The pole pieces are sometimes used to shape the magnetic field applied to the ferrites; this is done when we want to use magnets that are smaller or larger than the ferrites or have a different shape. For example, we may want to magnetize triangular ferrites having 3-inch altitudes using $1\frac{1}{2}$-inch diameter magnets, which could be accomplished with properly designed pole pieces. We will discuss the design of pole pieces in Section 6.4.

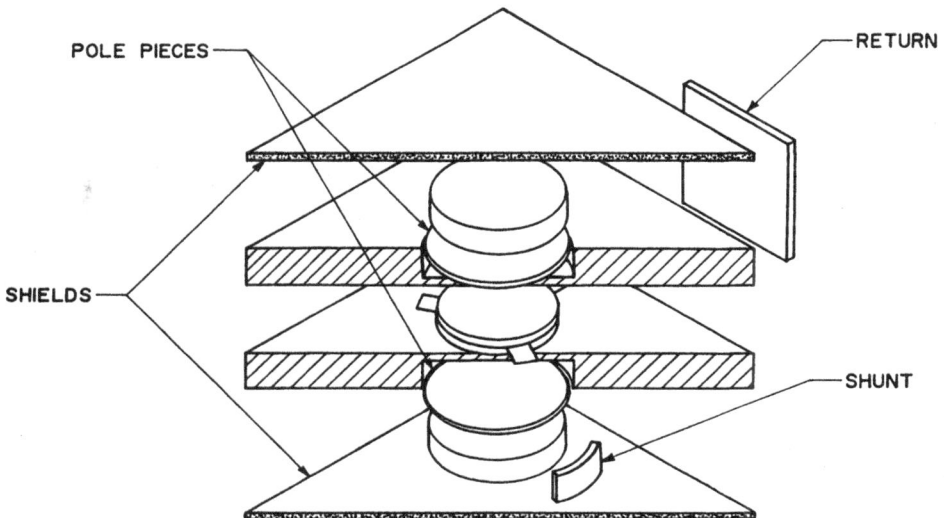

Fig. 6.3 Typical junction circulator magnetic circuit.

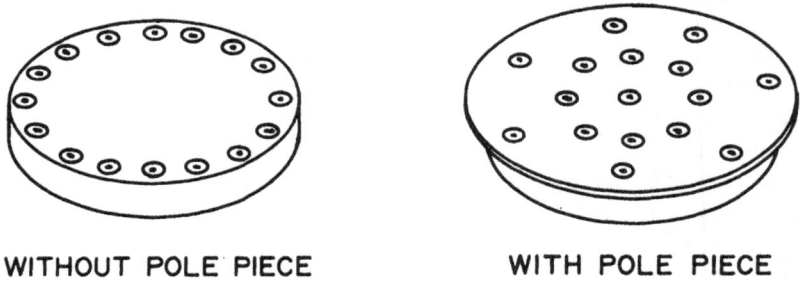

WITHOUT POLE PIECE   WITH POLE PIECE

Fig. 6.4 Effect of pole pieces on magnetic field.

Shunts are used to reduce the magnetic field supplied by the magnet to the ferrites. A shunt effectively short-circuits part of the magnetic flux from the magnet. Where it is undesirable or impossible to demagnetize a magnet, shunts are used. In some instances, we may need to skew the magnetic field to one side or another to compensate for magnetic anomalies; it can be done with shunts.

Returns serve the purpose of making the magnetic circuit more efficient and improving shielding. The return completes the magnetic circuit by returning the magnetic flux to its source, the magnet. Without returns, the outer faces of the magnets would be magnetically linked only by long flux paths through the air. These long paths are not efficient, and the magnets would have to be much larger to supply the needed field.

Shields not only provide magnetic shielding to prevent interaction between the circulator magnetic circuit and external magnetic fields or ferrous materials, but also improve the circuit efficiency by helping to return magnetic flux in much the same manner as returns. Section 6.2 is devoted to magnetic shielding.

It is important to understand the difference between a shunt and a shield or return. A shunt reduces the magnetic field intensity supplied to the ferrites. Shields and returns usually increase the field by providing a return path for the magnetic flux. These concepts are illustrated in Figure 6.5. The shunt effectively links the two poles of a magnet so that some fraction of the total magnetic flux provided by the magnet is short-circuited and therefore not supplied to the ferrites. The shields and returns link the outer poles of magnets on either side of the ferrites, thus providing a path of low resistance to magnetic flux. This path strengthens the magnetic field applied to the ferrites.

In this book we describe the design of junction circulator magnetic circuits; the magnetic design of other classes of circulators is substantially the same.

Before we can size magnets, we must know how much magnetic field is required by the ferrites. After we complete a preliminary electrical design of the circulator as described in Chapter 5, we know the approximate internal magnetic

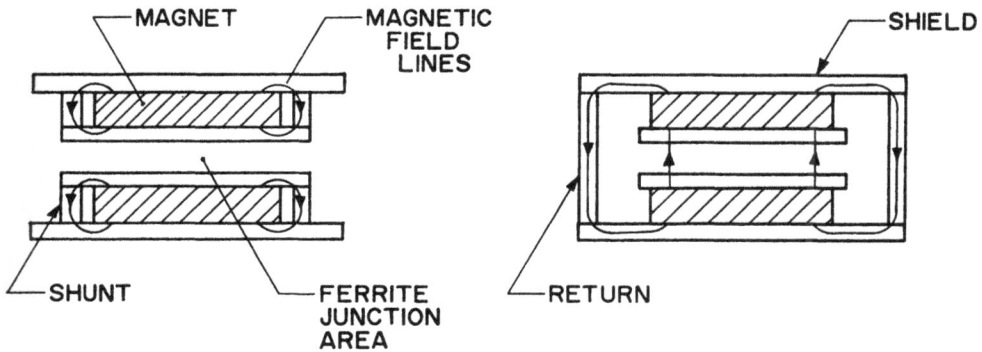

Fig. 6.5 Magnetic circuit components. Shunts short-circuit magnetic flux (left), returns and shields strengthen field in ferrite junction area (right).

field strength (H_{dc}) needed to bias the ferrites. Until now, we have only worked with the magnetic field inside the ferrite. The external magnetic field, H_{ext}, is different. It is the external magnetic field we use for the magnetic circuit calculations.

The external magnetic field differs from the internal field because of the creation of free magnetic poles [2] at the boundaries of the ferrites. We represent the effect of the free poles by demagnetizing fields that modify the internal magnetic fields. These modified internal fields are

$$h_x = h_{x\,ext} - N_x m_x \tag{6.2}$$

$$h_y = h_{y\,ext} - N_y m_y \tag{6.3}$$

$$H_{dc} = H_{ext} - N_z M_0 \tag{6.4}$$

where N_x, N_y, and N_z are demagnetizing factors. We substitute Equations (6.2) and (6.3) into (1.20) and (1.21), and derive

$$m_x = \chi(h_{x\,ext} - N_x m_x) - j\kappa(h_{y\,ext} - N_y m_y) \tag{6.5}$$

$$m_y = j\kappa(h_{x\,ext} - N_x m_x) + \chi(h_{y\,ext} - N_y m_y) \tag{6.6}$$

Solving Equations (6.5) and (6.6) simultaneously for m_x and m_y using determinants, we have for the system determinant:

$$D = \begin{vmatrix} 1 + \chi N_x & -j\kappa N_y \\ j\kappa N_x & 1 + \chi N_y \end{vmatrix} \tag{6.7}$$

Evaluating the determinant, we have

$$D = (1 + \chi N_x)(1 + \chi N_y) - \kappa^2 N_x N_y \tag{6.8}$$

Because the system determinant is the denominator in the expressions for m_x and m_y, and at resonance $m_x = m_y = \infty$ if we ignore losses, D must equal zero at resonance. We substitute Equations (1.25) and (1.27) into Equation (6.8), setting $\alpha = 0$, and derive

$$D = (\omega_0 + N_x \gamma M_0)(\omega_0 + N_y \gamma M_0) - \omega^2 \tag{6.9}$$

Setting $D = 0$, we have

$$\omega = \sqrt{(\omega_0 + N_x \gamma M_0)(\omega_0 + N_y \gamma M_0)} \tag{6.10}$$

Substituting Equation (1.7) into (6.10), we write

$$\omega = \gamma \sqrt{(H_{dc} + N_x M_0)(H_{dc} + N_y M_0)} \tag{6.11}$$

We bring back z-axis magnetic field by substitution of (6.4) into (6.11):

$$\frac{\omega}{\gamma} = \sqrt{[H_{ext} + (N_x - N_z)M_0][H_{ext} + (N_y - N_z)M_0]} \tag{6.12}$$

Insofar as we are writing equations for the resonance frequency, $\omega = \omega_0$ and we can again apply Equation (1.7):

$$H_{dc} = \sqrt{[H_{ext} + (N_x - N_z)M_0][H_{ext} + (N_y - N_z)M_0]} \tag{6.13}$$

This is Kittel's equation [3] relating the internal magnetic field strength to the external field.

An ellipsoid, depicted in Figure 6.6, has the property that when an external magnetic field is applied to it, the demagnetizing field is uniform throughout the solid. Because of this property, the ellipsoid is used as the model for determining the demagnetization factors. The demagnetizing factors N_x, N_y, and N_z are dependent on the ratios of the axes of the ellipsoid [4]. Let the axes of the ellipsoid be designated by a, b, and c. The demagnetizing factors are inversely proportional to the axes—when a dimension of the ellipsoid is small, the corresponding demagnetizing factor is large. The sum $N_x + N_y + N_z = 1$.

For a ferrite disk, we can approximate the demagnetization factors by letting the disk be an ellipsoid with axes $a = b = 1$ (because the disk is round, its x and y

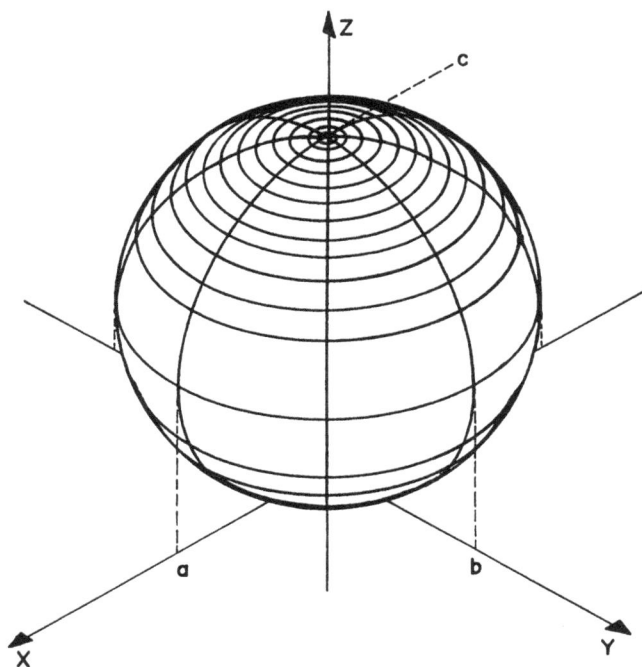

Fig. 6.6 The ellipsoid model for demagnetization factors.

dimensions are equal and arbitrarily 1) and $c = 10^{-10}$ (an arbitrarily small thickness). We find the axis ratios $b/a = 1$ and $c/a = 10^{-10}$. We set the demagnetization factors proportional to the reciprocals of the axis ratios:

$$N_x = \frac{N_x}{a} = N_x \tag{6.14}$$

$$N_y = \frac{a}{b} N_x = N_x \tag{6.15}$$

$$N_z = \frac{a}{c} N_x = 10^{10} N_x \tag{6.16}$$

We then equate the sum of the demagnetizing factors to 1:

$$N_x + N_x + 10^{10} N_x = 1$$

solving for N_x, N_y, and N_z, we have

$$N_x = 10^{-10}$$
$$N_y = 10^{-10}$$
$$N_z = 1$$

Because N_x and N_y are so much smaller than N_z, we set them equal to zero. For the disk shape, we find that $N_x = N_y = 0$ and $N_z = 1$.

The demagnetizing factors for other shapes can be calculated in the same manner. Table 6.1 summarizes the factors for the more important shapes.

Table 6.1

Shape	N_x	N_y	N_z	a	b	c
Thin disk	0	0	1	1	1	0
Slender rod	$\frac{1}{2}$	$\frac{1}{2}$	0	1	1	∞
Sphere	$\frac{1}{3}$	$\frac{1}{3}$	$\frac{1}{3}$	1	1	1
Thin slab	1	0	0	1	∞	∞

For the thin disk, used in junction circulators, the relationship between H_{dc} and H_{ext} becomes, by substitution of the demagnetizing factors into Equation (6.13):

$$H_{dc} = H_{ext} - M_0 \tag{6.17}$$

For the slender rod, used in Faraday rotators, we have

$$H_{dc} = H_{ext} + \frac{M_0}{2} \tag{6.18}$$

The sphere is used for testing ferrite materials because the internal field is very uniform. The expression for the sphere is

$$H_{dc} = H_{ext} \tag{6.19}$$

For differential phase shift circulators, field-displacement isolators, and resonance isolators utilizing rectangular ferrite slabs, the relationship between internal and external fields is

$$H_{dc} = \sqrt{H_{ext}(H_{ext} + M_0)} \tag{6.20}$$

We now have equations we can use to compute the amount of external magnetic field required, assuming we know the desired internal field. The ferrite material should always be saturated to avoid low-field losses and, as a result, we can replace the magnetization, M_0, with the ferrite material saturation magnetization. For a junction circulator with the ferrites just saturated ($H_{dc} = 0$), $H_{ext} = 4\pi M_s$.

The size of the magnets will depend not only on the field required by the ferrites but also on the reluctance of the entire magnetic circuit. We break the reluctance down into two components: the reluctance of the air gap (R_g) and the reluctance of the steel or iron path (R_i).

Reluctance is the magnetic equivalent of electrical resistance. It is also the reciprocal of permeance, which is analogous to electrical conductance.

The saturated ferrite material, aluminum, brass, copper, and other nonferrous metals all have permeabilities of essentially 1 (relative to air). The permeability of the iron or steel used in the magnetic circuit depends on the material and on the flux density in it. Figure 6.7 shows some B/H characteristics for ferrous metals used in circulators. If the magnetic intensity or flux density in the metal is known, the permeability can be read from these curves ($\mu = B/H$).

For the air gap, including ferrite and nonferrous metals,

$$R_g = \frac{L}{A} \tag{6.21}$$

and for the materials with $\mu > 1$,

$$R_i = \frac{L}{\mu A} \tag{6.22}$$

In both equations, L is the magnetic path length and A is the cross-sectional area of the path. Parallel paths can be combined in the same manner as parallel resistances, and series paths are simply added together [5].

The magnet length and area are calculated using [6]

$$L_m = \frac{B_g L_g}{H_d} \tag{6.23}$$

$$A_m = \frac{B_g A_g}{B_d} \tag{6.24}$$

B_g, the flux density in the air gap (also the flux density in the ferrite), is equal to H_{ext} because $\mu = 1$. A_g is the area of the air gap, or the area of the ferrite face. L_g, the gap length, can be calculated from

$$L_g = (R_g + R_i)A_g \tag{6.25}$$

B_d and H_d are the flux density and magnetic field strength of the magnet at the chosen operating point. Rearranging Equations (6.23) and (6.24), we derive

$$L_m = \frac{H_{ext}A_g(R_g + R_i)}{H_d} \tag{6.26}$$

$$A_m = \frac{H_{ext}A_g}{B_d} \tag{6.27}$$

These expressions can be used to size the magnets. The resulting length from (6.26) and (6.27) should be divided in half and one magnet placed on each side of the junction.

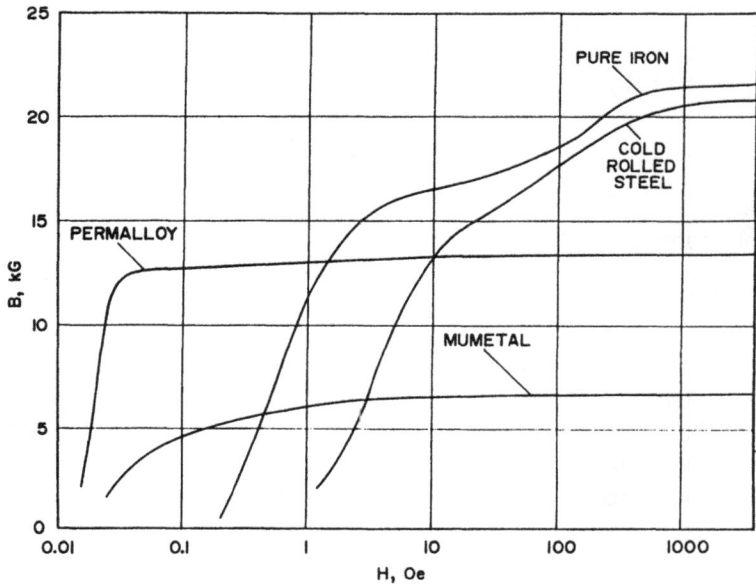

Fig. 6.7 Magnetization characteristics of ferrous metals used in circulator design.

Due to leakage flux, the lengths of the magnets may have to be increased by as much as 50% for large air gaps. Likewise, the magnet area may need to be increased by a factor of up to 20 to compensate for leakage flux. The calculation of leakage flux is quite complicated and lengthy, so it is best to determine the exact magnet dimensions empirically.

If it is necessary to recalculate the flux density in the ferrous portions of the magnetic circuit, we can use

$$B_i = \frac{H_{ext}A_g}{A_i} \tag{6.28}$$

where A_i is the cross-sectional area of the iron or steel.

To obtain maximum energy from the magnets, the operating point should be selected so that the product B_dH_d is maximized. This point is found using the demagnetization curve for the chosen material.

6.2 SHIELDING

Magnetic shielding of a circulator serves three purposes. First, it helps prevent magnetic fields external to the circulator from degrading its performance. Second, it protects other magnetically sensitive components in close proximity to the circulator from the circulator's field. Third, the shielding improves the circulator magnetic circuit efficiency by providing a low-reluctance path for flux that would otherwise be leakage flux.

The key to good magnetic shielding is to keep the magnetic flux provided by the circulator magnets inside the circulator and external flux outside the circulator. The shielding material should be chosen so that large amounts of magnetic flux can be conducted with very low reluctance. This means high permeability and high maximum flux density for the material. Unfortunately, high-permeability materials typically have lower saturation flux densities.

Low reluctance can be achieved either by high permeability or large cross section. Thus, if size is not important we can use shields of large cross section. Where low levels of flux are present, the high-permeability materials provide the best shielding for a given cross section.

It is important to keep the number of mechanical joints in the shielding material to a minimum to avoid magnetic discontinuities, which degrade shield effectiveness. In addition, smooth bends rather than abrupt ones make for better shielding.

When we compute the cross-sectional area of the shield material, we must consider not only the flux from the circulator magnets, but also flux from outside sources that may be present. If the external flux is high, it could easily saturate the shielding, after which the circulator would have no protection. We need to keep in mind that permanent magnets are not perfectly uniform flux sources. There will be regions where the flux density is higher (usually at the magnet edges). The shielding will have to accommodate flux densities in some areas larger than might be expected if the magnets provided a perfectly homogeneous field.

The flux density in the shielding material can be computed using Equation (6.28). The length of the shielding will, of course, be long enough to reach around the circulator. For optimum shielding, the greatest possible surface area of the circulator should be covered. In extreme cases, it may be necessary to use two layers of shielding, one inside the other.

The procedure for shield design is to first determine the allowable leakage flux from the circulator, and the amount of flux from any outside sources. Keep in mind that if ferrous materials are to be located near the circulator, these materials will provide a lower-reluctance path for leakage flux from the circulator, thus increasing the field applied to the ferrites. Next, we select a shield material and compute its minimum cross section based on the maximum flux density for the material. Remember to include the flux from the circulator magnets (Equation (6.28)) and external flux (B_i = flux lines/A_i). We then spread out this cross section over as much of the circulator surface as possible. If the shielding is designed after the magnets are sized, it will be necessary to resize them because of the lower-reluctance path.

6.3 TEMPERATURE COMPENSATION

To design a circulator with a wide operating temperature range, it is often necessary to apply temperature compensation to the magnetic circuit. Magnet manufacturers strive to produce magnets that provide a nearly constant flux regardless of temperature, but for most circulators we want a decrease in flux as temperature increases. Some devices may require an increase in flux with temperature.

Most compensating materials decrease in permeability as temperature increases. Thus, their position in the magnetic circuit will determine whether the circulator junction flux will increase or decrease with temperature. If the compensating material is placed in series with the magnetic path, flux decreases. If the material is in parallel, flux increases.

Ferrite compensating material in series is usually placed between the magnets and the pole pieces. Metallic compensators in series can be used as returns, and in parallel as magnet shunts.

Manufacturer's data for the compensating materials state how the permeability of the material varies with temperature for a specific magnetic field strength, as shown in Figure 6.8.

The first step in temperature compensation is to determine the required flux density as a function of temperature, either by calculation or testing. Then an appropriate compensating material can be selected. The magnetic operating point

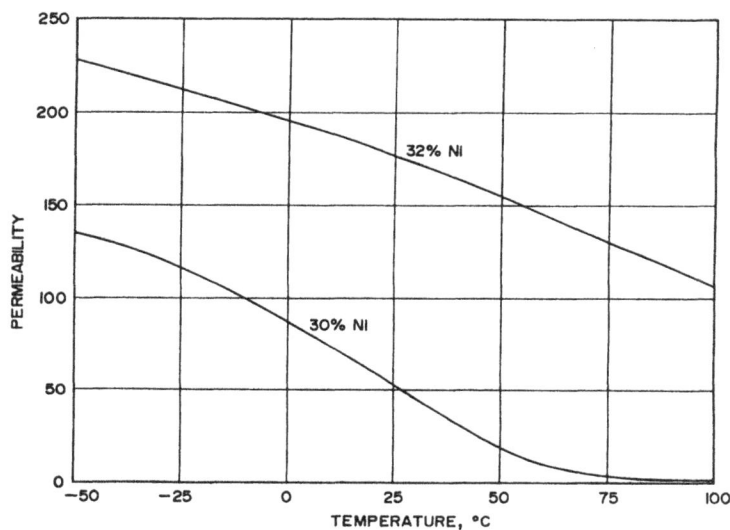

Fig. 6.8 Permeability *versus* temperature for typical iron-nickel alloys. $H = 50$ Oe.

should be set according to the manufacturer's data by adjusting the dimensions of the material to achieve the correct flux density. The flux density can be computed using Equation (6.28).

Circulators that contain discrete capacitors in their impedance-matching circuits, such as lumped-constant devices, can be temperature-compensated by selecting capacitors with different temperature coefficients.

Magnet and ferrite selection play an important part in the temperature stability of circulators, as do proper mechanical design and assembly.

6.4 COMPLETING THE CIRCUIT

In this section we will discuss pole pieces, the one remaining component of the magnetic circuit for which we have not presented design information.

The purpose of the pole piece is twofold: to homogenize the magnetic field and to shape the field. For low levels of magnetic field, such as those used for low-frequency below-resonance circulators, pole pieces are usually not needed. Where higher magnetic fields are utilized, as in above-resonance devices, pole pieces are a necessity, particularly with triangular magnets.

We cannot present exact designs for pole pieces, because the magnetic homogeneity of magnets varies dramatically for different magnets and shapes. For most triangular magnets used in above-resonance circulators, 16-gauge cold-rolled steel pole pieces are adequate. The pole pieces should have altitudes slightly larger than the magnet altitudes to be sure the leakage flux at the edge of the magnet is caught by the pole piece. In some cases, it may be necessary to make the pole piece the same size as the magnet, but the pole should never be smaller than the magnet unless the ferrite altitude is much smaller than the pole altitude.

Cold-rolled steel is used for pole pieces for several reasons. First, it has a permeability almost as high as that of pure iron at high flux densities. It is much less expensive than pure iron, which requires special processing to achieve high purity. Cold-rolled steel has a high saturation flux density, higher than the saturation levels of such high-permeability alloys as Mumetal and Permalloy. Lastly, cold-rolled steel typically has a smooth finish that does not require further machining, unlike hot-rolled steel, which typically has a scaled surface.

For higher-frequency below-resonance circulators with round magnets, pole pieces about half the thickness of the triangular ones previously discussed will serve. The diameter should be a little larger than the magnet diameter.

To change the shape of the magnetic field supplied by a magnet, we ideally want pole pieces machined to form smooth transitions in shape.

Figure 6.9 shows two pole-piece designs, one good and one bad. The bad pole-piece design uses a rather abrupt change in cross-sectional area, which leads to a large amount of leakage flux. The good design, on the other hand, accomplishes a lesser change in cross section but the absence of an abrupt dimensional change reduces the amount of leakage flux. Either design could work, but the good design would not require as large a magnet as the bad design.

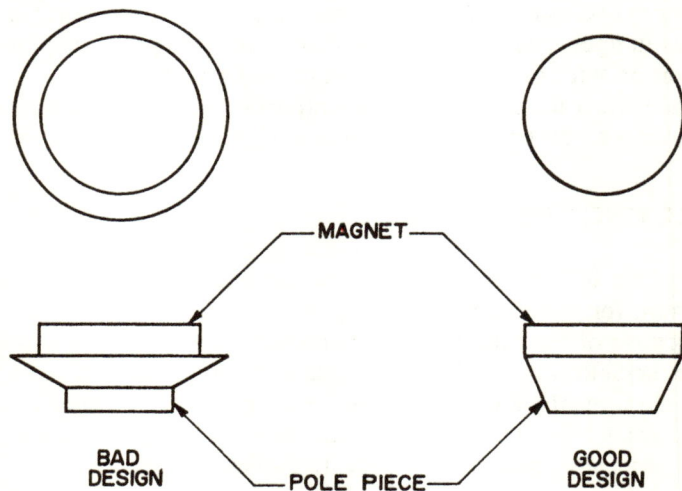

Fig. 6.9 Pole piece design: smooth transitions are best.

An optimum pole-piece design will minimize the amount of leakage flux caused by the pole piece. The pole piece should not be any thicker than necessary to homogenize the magnetic field so as not to cause excessive magnetic path reluctance. The thickness required will depend on the particular magnets used and the magnetic field intensity. In addition, the required degree of field homogeneity is a factor. Leakage flux minimization is done mainly by avoiding abrupt dimensional changes.

Pole-piece performance is very difficult to analyze quantitatively because of the difficulty in analyzing the leakage flux and the exact shape of the magnetic field provided by a magnet. Hence, pole-piece designs are best derived empirically. We should also stress that pole-piece design is not critical—many topographies are usable.

REFERENCES

1.   Betts, F., D. H. Temme, and J. A. Weiss, "A Switching Circulator: S-Band; Stripline; Remanent; 15 Kilowatts; 10 Microseconds; Temperature-Stable," *IEEE Transactions on Microwave Theory and Techniques,* December 1966, pp. 665–669.
2.   Clarricoats, P. J. B., *Microwave Ferrites* (New York: John Wiley and Sons, 1961).
3.   Kittel, C., "On the Theory of Ferromagnetic Resonance Absorption," *Physical Review*, January 15, 1948, pp. 155–161.
4.   Soohoo, R. F., *Microwave Magnetics* (New York: Harper and Row, 1985).
5.   *Advanced Electronics,* AE-4 (Los Angeles: National Technical Schools).
6.   Magnet Catalog P5A (Hicksville, NY: Permag Corp., 1986).

Chapter 7
Mechanical Design

7.1 COAXIAL JUNCTION CIRCULATORS

The final stage in the circulator design process is the mechanical design. In this stage, we package the ferrites, magnets, connectors, and other components, forming a microwave device that meets specifications. The specifications for the circulator may include not only electrical parameters and environmental factors but also cost. The type of mechanical design used will be dependent on the available manufacturing facilities as well as specifications.

Most coaxial junction circulators are based on stripline techniques. There are five basic packaging schemes for stripline circuits [1]. The first is flat-plate construction. This is an inexpensive and simple technique where two flat metal plates are placed outside the center conductor and dielectric material (or ferrite), one on each side. The plate thickness is chosen to provide rigidity to the circuit and to allow screw holes for connector mounting at the edges. However, it is not imperative that the connectors be mounted at the edges. Broad-wall launching could be used.

The plates are usually held together with screws; we use metal spacers between the plates if the dielectric is not firm enough to prevent bending the plates when the screws are tightened. A drawback of this construction is that the edges of the device, between the ground plates, must be sealed to prevent RF radiation and to protect the circuitry from the environment and foreign objects. Figure 7.1 illustrates the flat-plate construction.

The second packaging scheme we will discuss is the bonded-substrate construction, shown in Figure 7.2. Dielectric materials are available that have aluminum plates already bonded to them. This scheme reduces the amount of machining required, but the plates are usually too thin to have connectors screwed to them. Consequently, we must either use special connectors or manufacture

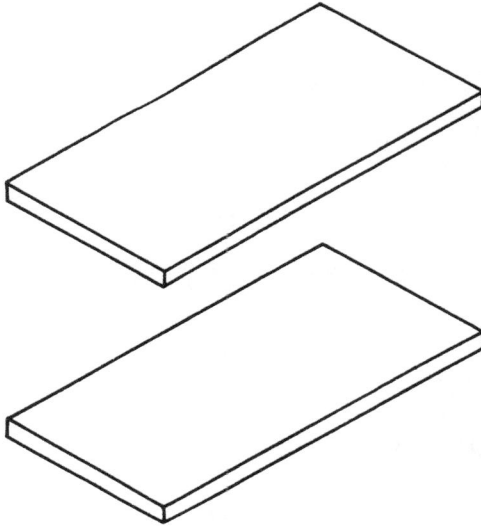

Fig. 7.1 Flat-plate construction.

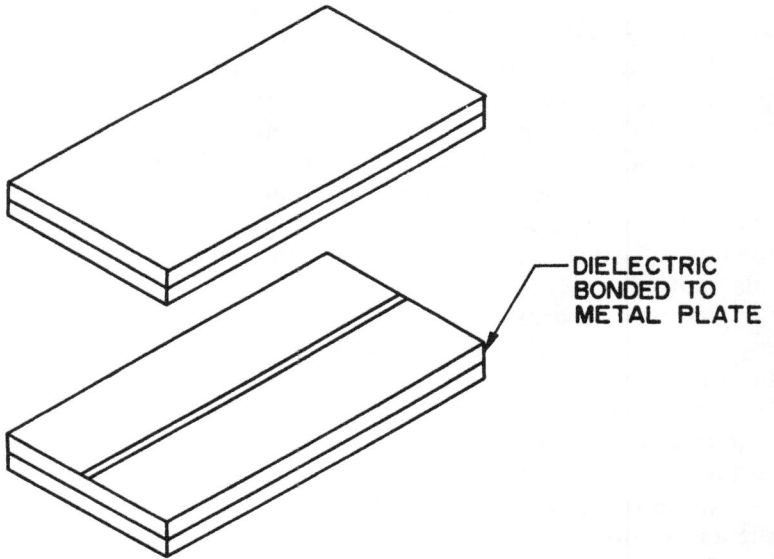

DIELECTRIC
BONDED TO
METAL PLATE

Fig. 7.2 Bonded-substrate construction.

blocks for connector mounting. The bonded-substrate construction also requires that we seal the edges of the device.

Figure 7.3 depicts another stripline technique that utilizes metal plates with channels cut into them. The edges of a device constructed using this technique do not usually require additional sealing. A disadvantage of this scheme is that the dielectric material must be cut to fit into the channels, unless air is the dielectric. In this case, the center conductor must be properly supported. The plates can be channeled by various means, including milling, electrical-discharge machining, drawing, and high-energy-rate forming [2].

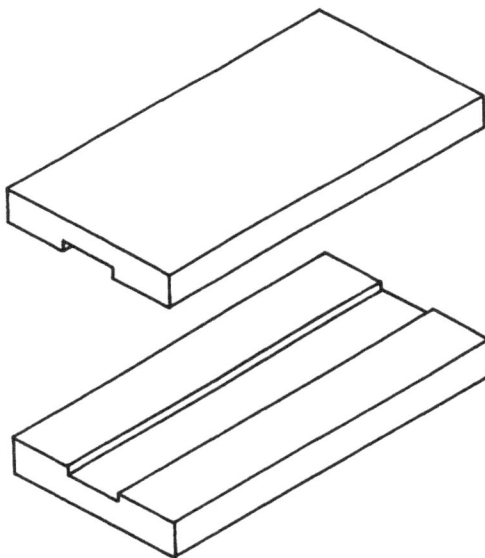

Fig. 7.3 Channeled-plate construction.

Box-and-cover construction, shown in Figure 7.4, is very similar to the channeled-plate construction. The box-and-cover technique does not require that both plates be channeled; only the box plate is channeled. This construction reduces the amount of machining to produce a pair of plates, but different machine setups may be needed to produce each plate. In addition, more time is spent drawing prints for a box and cover than would be spent on prints for two channeled plates, because the two channeled plates could be designed as symmetrical mirror images of one another. Symmetrical mirror-image parts can be machined using the same setup. The only necessary difference between the parts is related

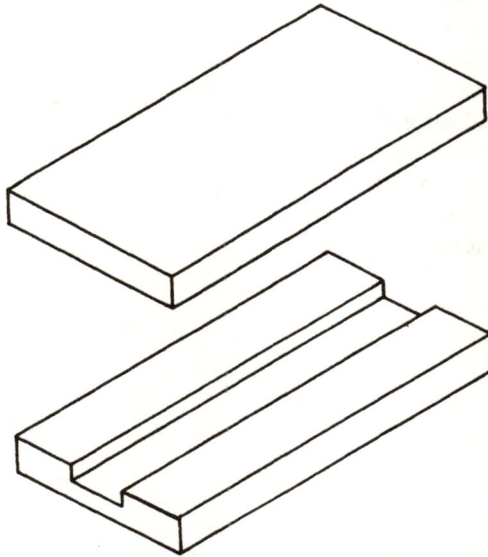

Fig. 7.4 Box-and-cover construction.

to the screws that hold them together—one will have tapped (threaded) screw holes and the other will have clearance holes.

The decision between the channeled-plate construction and the box-and-cover construction should be made based on the quantity of parts to be manufactured.

The fifth stripline packaging scheme is the bonded-stripline construction. This technique uses dielectric sheets with copper ground planes on the outside. The circuit is etched on the other side of one or both dielectric sheets. The sheets are held together by a resin bond formed by heat and pressure [3]. The resulting assembly is highly resistant to humidity and salt spray and is very lightweight, but cannot be repaired and does not offer rigid support to components. Bonded stripline techniques are unsuitable for circulators because of the lack of support given to the ferrites.

The metal plates in the first four packaging schemes can be replaced by metalized plastic parts. Such parts are being used more frequently where weight is an important parameter.

Of the five stripline packaging techniques, three are most suitable for circulators. As previously mentioned, bonded stripline does not offer support to the ferrites. The bonded-substrate configuration does not lend itself to circulator design because we would need ferrite slabs with the ground planes bonded to them, which would lead to difficulties in machining the combination and to wasted ferrite. We are left with the box-and-cover construction, the channeled-plate technique, and the flat-plate construction.

An important difference between the flat-plate construction and the other two is that the edges of the flat-plate device must be sealed. An advantage of the flat-plate scheme is that intimate contact between the ferrites and the ground plane is achieved by tightening the screws to the correct torque. Good contact between the ground plane and the ferrite can be obtained with the other two constructions only by imposing tight dimensional tolerances.

Imperfect contact between the ferrites and the ground planes can cause undesired resonances and generally poor electrical performance. It is therefore imperative that the ground planes be very flat and very smooth. Circulator housings (also called bodies) that have channels are usually dimensioned so that there is a slight amount of compression applied to the ferrite junction, ensuring intimate contact between ferrite and ground plane. If the depths of the channels in the housing are relied on to set up the appropriate compression, the tolerance on the ground-plane spacing must be tight: ±0.001 inch or so. Figure 7.5 illustrates the effect of incorrect ground-plane spacing. When the spacing is too large, the ferrites are not in contact with the ground planes. When the spacing is too small, the ground plane bows up at the center of the ferrite, not making good contact. In extreme cases, the ferrites may be broken when the circulator is assembled. The ground plane may be permanently bent up in the center so that it will never make good contact with the ferrite. A good amount of interference (compression) with which to design is about 0.001 inch. We should remember that ferrites are hard and are not compressible as are many dielectric materials.

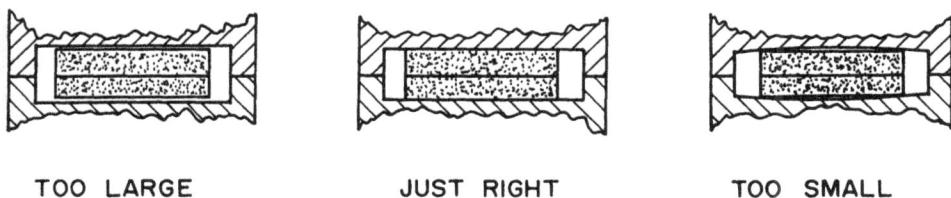

TOO LARGE   JUST RIGHT   TOO SMALL

Fig. 7.5 Effects of incorrect ground-plane spacing.

Compression can be achieved with channeled circulator housings using two methods that do not rely on the channel depth to set up the ground-plane spacing. One method is to set the channel depth slightly less (0.010 to 0.020 inch) than the depth required for ferrite contact. Then, the screws that hold the housing halves together can be tightened to a torque that provides for good ferrite contact. This technique defeats the purpose of the channels and the circulator must be sealed.

The second method is to place a compressible material such as rubber or neoprene between the ferrites and the ground planes. Ground continuity must be maintained by laying thin metallic foil between the ferrites and the compressible material. When the housing halves are assembled, the compressible material provides a force, pushing the ground foil against the ferrite. This method is shown in Figure 7.6. The rubber or neoprene pieces also may serve as shock mounts for the ferrites. The disadvantage of this technique is that the ground will not be perfectly flat and smooth because the thickness of the compressible material may vary and the foil may be wrinkled. The compressible material will degrade with age.

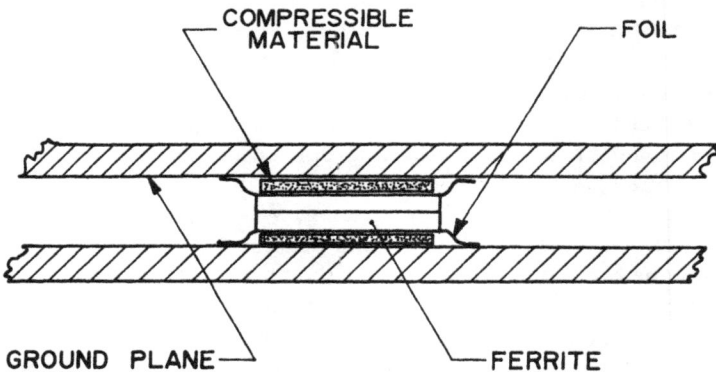

Fig. 7.6 Ferrite junction compression achieved using compressible material.

The thickness of the metal between the ferrite and the magnet or pole piece (web thickness) should be thick enough so that it will not flex much when the circulator is assembled. There are practical limits to how thin the web can be made from a machining point of view as well. Circulator housings are often machined from solid blocks of aluminum. If a very thin (flexible) web is specified, chances are good that the milling cutter will accidentally punch right through the web, destroying the part; it could happen if the cutter is pulling up on the web and the web is flexing so that it is cut when it should not be.

It is desirable to have a thin web from a magnetic point of view because this will reduce the air gap in the magnetic circuit, thus reducing the magnet size. Typically, webs can easily be made as thin as 0.030 inch. However, webs on the order of 0.100 inch thick are desirable electrically because they usually provide flatter ground planes.

Thermal considerations also influence the selection of web thickness. If we know the Curie temperature for the ferrite material used and the insertion loss of the ferrite junction, either by calculation or measurement, we can compute the temperature the ground plane must be to avoid severe electrical performance degradation due to RF heating of the ferrites. The thermal conductivities [4] of some materials commonly used in circulators are given in Table 7.1.

Table 7.1

Material	Thermal Conductivity (W/in °C)
Aluminum	5.5
Brass	3.1
Copper	10.0
Steel	1.7
Air	0.0007
Alumina	0.7
Beryllia	5.0
Epoxy (conductive)	0.02
Heat sink compound	0.01
Teflon	0.005
Ferrite	0.16

Making the safe assumption that all RF heating takes place at the surface of the ferrite, we can apply the following formula to compute the temperature drop across the ferrite:

$$T = \frac{Qd}{kA} \tag{7.1}$$

where Q is the power dissipated in the circulator junction, calculated from the average input power and the insertion loss in dB, using

$$Q = P_{in}(1 - 10^{(-\text{loss}/10)}) \tag{7.2}$$

and d is the ferrite thickness, k is the ferrite thermal conductivity from Table 7.1, and A is the cross-sectional area of the ferrite. Because the power is not evenly distributed on the surface of the ferrite, we need to apply a factor to compensate for this. If we use a factor of 2 (which implies that the power is dissipated in only half of the ferrite), and remember that there are actually two thermal paths, one for each ferrite and ground plane, we find that Equation (7.1) already includes the compensation factor.

Once we know the temperature drop across the ferrite, we subtract this from the ferrite Curie temperature to find the maximum ground-plane temperature. If this temperature is significantly below the ambient air temperature, methods of cooling the circulator or decreasing the power density in the ferrite must be investigated.

If the average power level is very high and the ground-plane temperature must be held very low, water cooling is indicated. Water cooling is accomplished by placing a water jacket between the ferrite and magnet or pole piece, in place of the web. The temperature rise of the water jacket above the water temperature per watt of power dissipation is given by [4]

$$\frac{T}{Q} = \frac{0.72\ A^{0.4}}{F^{0.8}L} \tag{7.3}$$

where A is the cross-sectional area of a cooling duct of circular cross section, F is the water flow in gallons/minute, and L is the length of the cooling duct. The assumption is made that the flow rate is high enough so that the flow is turbulent, that is,

$$\frac{F}{A^{0.5}} \gg 0.69 \tag{7.4}$$

A less drastic cooling measure is forced-air cooling. Before we can apply this cooling method, the power dissipated in the ferrite has to be conducted to the surface of the circulator. To calculate how much lower than the ground-plane temperature the circulator surface must be held, we apply Equation (7.1) to all segments of the thermal path from the ground plane to the surface, summing the temperature drops. The length of each segment is d, the cross-sectional area is A, and the appropriate thermal conductivity (k) is used. Other equations can be used for special geometries; they are beyond the scope of this book.

The temperature rise of a forced-air(fan)-cooled heat sink per watt of power dissipation is approximated by [4]

$$\frac{T}{Q} = \frac{140\ w}{n^{0.2}z^{0.2}F^{0.8}L} \tag{7.5}$$

where w is the heat-sink fin spacing, n is the number of fins, z is the fin height, F is the air flow in ft³/min, and L is the length of the heat sink.

The approximate amount of power that can be transferred to the ambient air by a natural-convection heat sink is shown in Figure 7.7 as a function of the temperature difference between the heat sink and the ambient air.

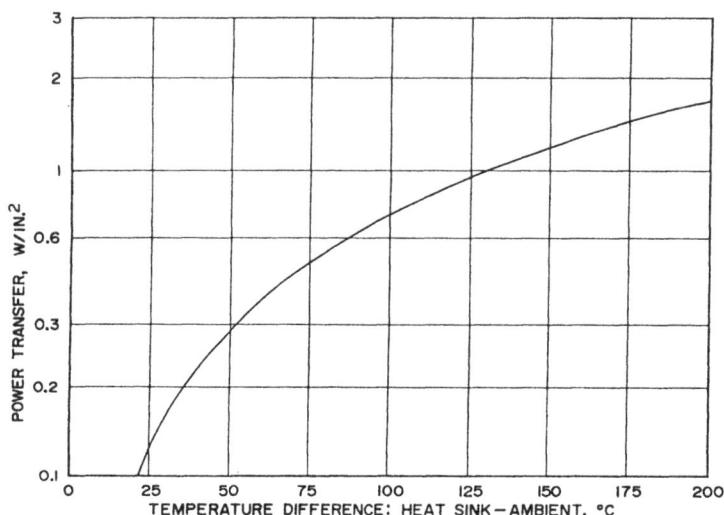

Fig. 7.7 Power transferred by a natural-convection heat sink *versus* temperature rise [4].

Conduction cooling is yet another cooling method. When conduction cooling is specified, maximum circulator case temperature must also be specified. Without a maximum temperature, the cooling system designer will not be able to do his or her job properly. It is helpful to the user to know the approximate amount of power dissipated in the circulator.

To determine the proper type of cooling for an isolator, we need to consider the power dissipated in the internal termination. This power depends on the VSWR of the load connected to the isolator output.

We can use the conduction formula (7.1) to compute the power-handling capacity of a strip transmission line. To do this, we find the insertion loss of the line per unit length, then find the power dissipated in the line. Knowing the maximum allowable temperature of the center conductor or dielectric, we can plug the appropriate values into Equation (7.1) and come up with a maximum average power, Q.

For both the box-and-cover construction and channeled-plate construction, we need to determine the widths of the channels and the size of the center cavity. These dimensions are indicated in Figure 7.8. Note that the exterior shape of the

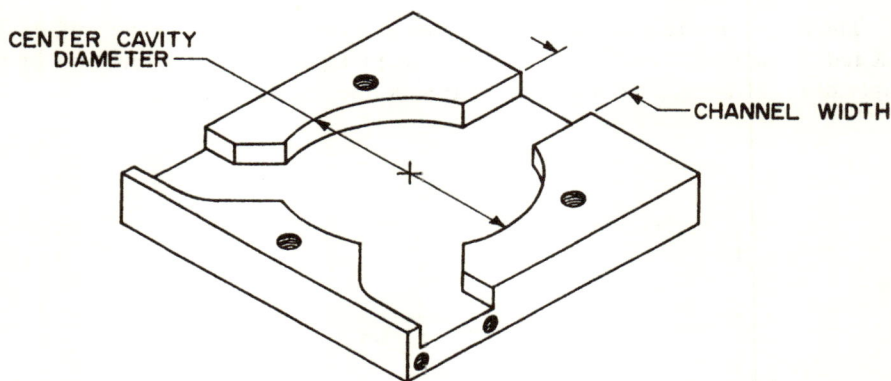

Fig. 7.8 Dimensions of box-and-cover or channeled-plate housing half.

circulator housing shown in Figure 7.8 is rectangular. The shape could also be round, triangular, or hexagonal. The rectangular shape is normally used because it is easily integrated into a system. Shapes with three-way symmetry can be difficult to mount without wasting space.

The size of the center cavity, which will be either round or triangular depending on the ferrite shape, is determined from the size of the ferrite and any dielectric material that may be attached to the ferrite. Where impedance transformers are used, dielectric material may be either located in the channels connecting the center cavity and the connectors or immediately adjacent to the ferrite in the center cavity. If two quarter-wavelength transformers are used (a typical construction for octave-bandwidth circulators), and these transformers are located adjacent to the ferrite, the center cavity diameter will be increased by approximately a wavelength, as shown in Figure 7.9. A cavity of this size will support circular resonance modes, which may result in high insertion loss at some point in the band. These circular modes can be suppressed by strategic location of absorbing material such as polyiron. The dielectric material can be cut into slabs and placed in the channels. If we use this technique, we do not have the problem of circular modes, but more machining is required on the dielectrics.

If the dielectric ring illustrated in Figure 7.9 is used, the center cavity should be dimensioned so that the dielectric just fits, allowing for dimensional tolerances. When the cavity size is based only on the ferrite diameter or altitude, we dimension the cavity so that the walls are far enough away from the ferrite to prevent their being part of the circuit. Usually the walls have little influence on the ferrites if the space between a wall and the ferrite is approximately equal to the ground-plane spacing. Typically, we can make the spacing larger in above-resonance circulators because the μ_{eff} is higher and the ferrite altitude or radius is a smaller

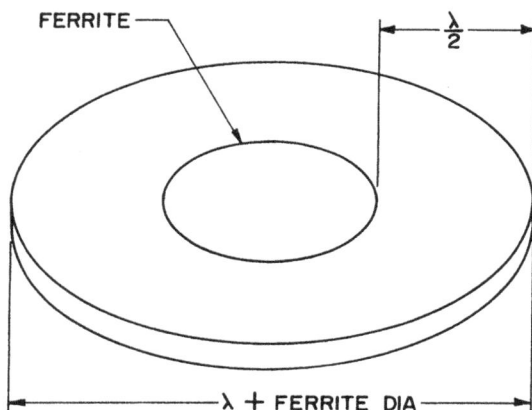

Fig. 7.9 Ferrite disk in dielectric ring for dual quarter-wavelength transformers at each port.

portion of a wavelength. Therefore, the cavity can be larger before it becomes a large enough fraction of a wavelength to resonate.

We select the channel widths such that they will not affect the characteristic impedance of the striplines. There is certainly no significant change in Z_0 if the walls are two line widths from the center conductor, and the spacing can be as little as one-half the line width depending on the impedance. Stripline design data are available that include the effect of the walls, if it is necessary to make the channels very narrow.

The channels should be located 120 degrees apart or the circulator will not be symmetrical electrically. It is also desirable to make the channels straight, if possible, to avoid discontinuities.

The flat-plate construction has neither channels nor center cavity. Because the inside of a circulator made using this method of construction is completely open, several different types of resonances can occur. To help eliminate these parasitic resonances and couplings, we can place screws around the ferrites and along the sides of the striplines, spaced an eighth of a wavelength or less apart. Some resonances may not show up until the circulator is sealed. Absorbing material can also be useful in the reduction of resonances. The exact location of this material is best determined empirically.

In any type of stripline circulator, it may be necessary to make transitions in stripline characteristic impedance. These transitions could be between transformer sections or between a transformer and a 50 Ω section. Step transitions are the simplest, but create the worst discontinuity. We can select one of two other commonly used techniques: the triangular compensation method illustrated in Figure 7.10, or a contour method [5].

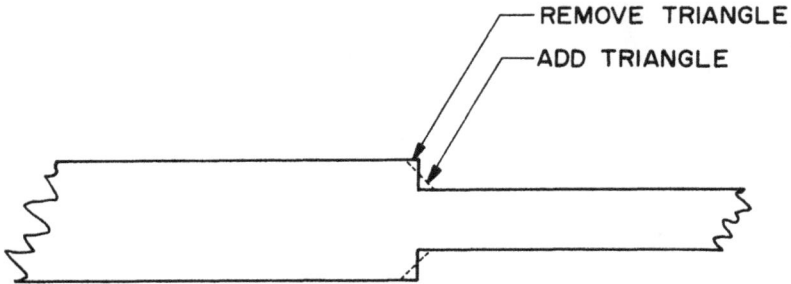

Fig. 7.10 Stripline-transition triangular compensation technique.

A stripline step transition discontinuity can be modeled using the equivalent circuit shown in Figure 7.11(a). This is an approximate equivalent circuit, where the inductance is given by

$$L = \frac{60\pi b}{\omega\lambda} \ln \left[\csc \left(\frac{\pi Z_{t2}}{2Z_{t1}} \right) \right]$$

where b is the ground-plane spacing and Z_{t1} and Z_{t2} are the low and high characteristic impedances of the two striplines, respectively.

If the inductance added by the step transition is unacceptable, we can use the triangular compensation method. In this method, we attempt to minimize the discontinuity equivalent inductance by removing and adding triangular sections as shown in Figure 7.10. The amount to be removed or added can be determined by breaking up the transition into small steps and computing the discontinuity inductances for each of the steps. If the length of these steps for analysis is much less than a wavelength, the analysis will be fairly accurate. Another way to determine how large the triangles to be removed and added should be is to use an experimental (cut-and-try) method.

A more accurate stripline-transition discontinuity model is shown in Figure 7.11(b). This model could be applied along with a technique to maintain a constant characteristic impedance up to the actual transition, resulting in a contoured transition. The details of such a compensation technique are beyond the scope of this book, but [5] at the end of this chapter describes a computer program to help reduce discontinuities.

We need to seal the finished circulator, especially if it is of flat-plate construction, to prevent RF radiation and to protect the inside of the circulator from humidity, rain, salt spray, and dust. One of the most frequently used sealing schemes for circulators is to epoxy edge covers (of aluminum or steel) in place. Epoxy is also used to seal joints. The disadvantage of epoxy is that it is difficult to repair the circulator once it is sealed.

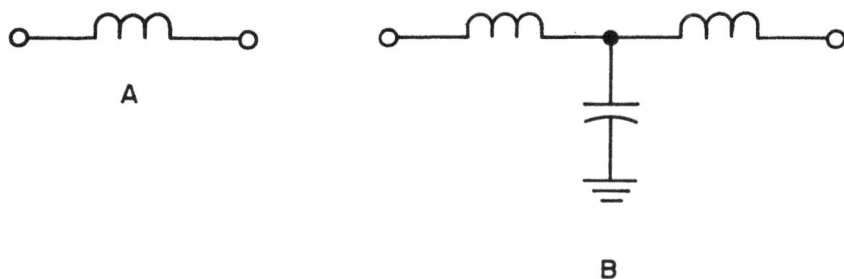

Fig. 7.11 Stripline-transition discontinuity equivalent circuits.

We can use gaskets, either ordinary or of the RFI variety, to seal the circulator edges. This is a good method if the circulator is to be pressurized to increase the power-handling capacity and to reduce RF radiation.

Another sealing method is the use of metal tape with adhesive. We use this tape to seal the edges and joints of the circulator.

Where hermetic sealing is required, we use glass-bead connector interfaces and solder or weld the housing together. If the soldering or welding is done properly, the circulator will not be overheated, because only the immediate area of the joint will be heated.

The magnets are typically placed in wells so that the plates can be thicker than the web thickness between the magnet and ferrite. These magnet wells are positioned so that the ferrites are centered beneath the magnet. Some engineers believe that they can achieve better performance from the circulator by deliberately skewing the magnetic field off to one side. This technique may optimize one parameter but will degrade others. At best, it is a compromise of circulator performance.

The magnet wells should be sized so that the pole piece and magnet will easily fit inside, but not so large that the performance of the circulator changes because the magnet moves around. The bottom of the magnet well must be parallel to the ground plane, or field asymmetry will result.

Where magnet wells are not used, as in the flat-plate construction, the magnets must be held in position by epoxy or mechanical means.

Finishing is an aspect of circulator design we do not want to overlook. Where dissimilar metals such as steel and aluminum are in contact with one another, it is important to plate one or both metals to make them compatible. Failure to do so results in galvanic action and erosion of one of the metals. Steel should not be left bare in any case because it rusts. Either zinc or cadmium plating can be used to protect the steel. Aluminum is frequently finished with a chemical-conversion coating such as iridite.

The outside finish on a circulator will depend on the severity of the environment. Finishes range from attractive plating such as chrome or nickel to epoxy coatings and paint.

7.2 LUMPED-CONSTANT CIRCULATORS

The mechanical design of lumped-element circulators differs from that of stripline junction circulators. The operating frequency range is usually lower; resonances are not normally a problem. The location of magnets relative to the ferrites is very similar to the junction circulator magnetic circuit, but in the lumped-element circulator there must be space for impedance-matching capacitors and inductors between the ferrites and the connectors. A typical lumped-element circulator housing is shown in Figure 7.12.

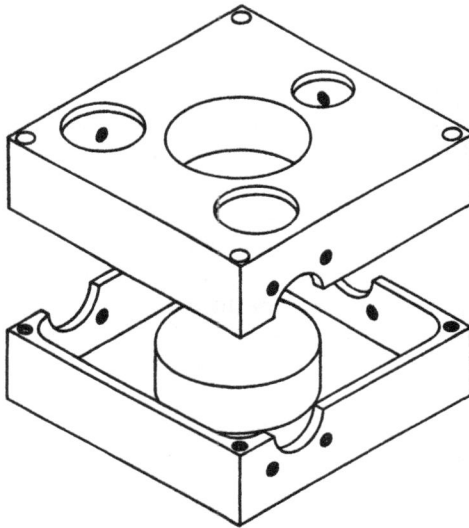

Fig. 7.12 A typical lumped-element circulator housing.

The type and size of matching circuit components will determine the amount of space needed in the housing and the method of access to the components for tuning. For low-power circulators, tuning can be accomplished using trimmer capacitors, which can be mounted so that adjustment is possible without access holes. For higher-power circulators, it may be necessary to design access holes so that the matching circuit components can be adjusted without opening the circulator; this is necessary because the ferrite junction will behave in quite a different manner when the upper ground plane and magnet are removed.

After the component values for the impedance-matching circuits have been determined by evaluation of the ferrite junction itself and application of appropriate network synthesis methods, we proceed with the mechanical design of the inductors and capacitors. At low power levels, we can use off-the-shelf components. High power levels demand that we design our own components to ensure adequate electrical performance.

The first step in the inductor design procedure [6] is to compute the expected RF current. From this, we calculate the necessary wire diameter using

$$d = \frac{I_{RF}(f)^{1/4}}{500} \quad \text{(inches)} \tag{7.6}$$

where f is the frequency in MHz and I_{RF} is the RF current in amperes. The suggested spacing between turns of the coil [6] is one wire diameter. Next, we compute the coil diameter for optimum Q:

$$D = 8.9(Ld^2)^{1/3} \tag{7.7}$$

where L is the required inductance in microhenries. If the diameter is impractically large, we reduce the diameter to a practical value and calculate the number of turns:

$$N = \frac{40\,Ld}{D^2} + \left[\left(\frac{40\,Ld}{D^2}\right)^2 + \frac{18\,L}{D}\right]^{1/2} \tag{7.8}$$

Lead inductance and stray capacitance depend on how the coil and its leads are positioned, so it is best that coil adjustments to compensate for these factors be done experimentally. Large coils need to be supported. Teflon rods serve well as coil supports. If there is a danger that the coil or leads will short to each other or ground, Teflon tubing should be placed over the wire.

To design a capacitor, we first compute the RF voltage across the capacitor, then select a dielectric material and thickness that will safely stand off this voltage. Teflon is a good choice because of its high dielectric strength and low loss factor. For a parallel-plate (or parallel-disk) capacitor, the capacitance is given by [7]

$$C = \frac{0.225\varepsilon A}{t} \quad \text{(pF)} \tag{7.9}$$

where A and t are the plate area and dielectric thickness, in inches respectively. A cylindrical-coaxial capacitor may have a more convenient shape than the parallel-plate capacitor. Its capacitance is [7]

$$C = \frac{0.614\varepsilon}{\log_{10}(b/a)} \quad \text{(pF/inch)} \tag{7.10}$$

where b and a are the outer and inner diameters of the dielectric in inches and ε is the relative permittivity of the dielectric.

Some lumped-element circulators use a series-resonant circuit from the common node of the three mutually coupled inductors wrapped around the ferrites to ground. In these circulators, the ferrites cannot be in direct contact with the ground planes and dielectric material must be placed in the gap. In low-power circulators, a soft material such as Teflon may be used, but for higher powers, the dielectric must be a good heat conductor in order to keep the ferrites cool. In this case, alumina and beryllia are good materials to use.

Many of the design concepts presented in Section 7.1 for coaxial junction circulators, such as those for cooling, sealing, compression, and finishing, also apply to lumped-element circulator design. As mentioned previously, the primary differences lie in the operating frequency (parasitics are of less concern with regard to lumped circulators) space required for discrete capacitors and inductors in the lumped circulator, and provisions for tuning.

7.3 WAVEGUIDE CIRCULATORS

Two types of waveguide circulators have been presented in this book: the waveguide junction circulator and the differential phase shift circulator. Many of the concepts we discussed in Section 7.1 with regard to coaxial junction circulators also apply to waveguide junction circulators. The waveguide versions are constructed using the channeled-plate or box-and-cover methods. For large (low-frequency) circulators, sections of waveguide may be welded or brazed together to form a Y or star, because machining the circulator housing from solid blocks of metal would not be economical.

The remarks in Section 7.1 about web thickness and cooling are applicable here as well. An important difference in the mechanical design of waveguide junction circulators is in the quality of machining. If the waveguide is machined into a metal plate, as in the channeled-plate and box-and-cover constructions, the dimensions of the guide (depth and width) must be very accurate and meet the specifications for the particular waveguide size. The maximum allowable perpendicularity [8] deviation of the walls is typically 0.5 degree. Military specification MIL-W-85G states that the maximum allowable bow of the narrow wall is 0.00042 inch/inch and 0.00083 inch/inch for the broad wall. The waveguide cannot be twisted more than 0.083 degree per inch of length. The inside surface roughness should be 32 microinches or better and any scratches present must not be more than 0.001 inch deep. Circulators designed to be pressurized must be tightly sealed to prevent leakage. The mating surfaces of the waveguide flanges must be very smooth and flat, and the bolt holes should be the correct size and in the correct locations. The type of waveguide flange to be used should be included in the circulator specifications.

It is doubtful that the circulator performance would be severely degraded if we did not adhere to the preceding specifications. It is true, however, that the machining of waveguide circulators, especially those for high frequencies, is more critical than coaxial circulator housing machining.

The ferrites in a waveguide junction circulator are typically located as shown earlier in Figure 5.7. For geometries (b) and (c), we mount the ferrites to the waveguide wall using adhesive or solder. The adhesive is an epoxy; conductive epoxy will provide a good ground plane at the ferrite face. The ferrites can be metalized on one face and then soldered inside the waveguide. This provides a good thermal and electrical path if there are no air pockets in the solder. We can also hold the ferrites in place with nonmetallic clamps or screws if the thermal and electrical properties of the interface are not critical.

The mounting of the ferrite post in Figure 5.7(a) is somewhat simpler because the entire space between the waveguide walls is filled. If we apply pressure to the waveguide walls (webs) from the outside, the dielectric spacers and ferrite post are held in place without using solder or adhesive. If the circulator must withstand shock and vibration, some method of captivating the ferrite and dielectric is necessary.

The mechanical design of differential phase shift circulators includes the design of a magic tee and quadrature coupler and the design of differential phase shifters. Design data for couplers and tees are readily available elsewhere, so we will not discuss their design here. Purchased tees and couplers could be used in the construction of differential phase shift circulators.

Mechanically, a differential phase shifter is little more than a section of waveguide with a ferrite slab or slabs inside and magnet outside. The external magnetic circuit, which may be just the magnet itself, is usually C-shaped, the ends of the C contacting the waveguide walls as shown earlier in Figure 2.10.

When the ferrite slab is positioned as previously shown in Figure 5.23, it can be held in place by filling the guide with a dielectric foam. Foams are available that have dielectric constants near unity, so they will not significantly affect the operation of the phaser. Sometimes dielectric slabs are placed alongside the ferrite slab; they help to hold it in place.

When the ferrite slabs are placed against the broad wall of the waveguide as shown in Figure 2.13, they are usually held in place with adhesive as used for junction circulators.

A final mechanical consideration of differential phase shift circulators is to make sure we check the dimensions and orientations of the tee, coupler, phasers, and dummy load (in the case of an isolator) to be sure all the components fit together properly.

7.4 RESONANCE ISOLATORS

The mechanical design of waveguide resonance isolators is very similar to the design of differential phase shifters. Dielectric slabs are normally placed against the ferrite face nearest the center of the guide. The difference between the resonance isolator and the differential phaser is the magnetic operating point, the resonance isolator requiring a stronger magnetic field. In addition, reverse power is absorbed in the ferrite of a resonance isolator, so more consideration needs to be given to the cooling of the ferrite slabs, particularly at high power levels. The thermal analysis methods presented in Section 7.1 can be applied here.

Coaxial resonance isolators are constructed from straight sections of coaxial transmission line, which are not difficult to design. Perhaps the most difficult aspect of resonance isolator manufacture is the machining of the dielectric insert.

We do not present detailed design information for the coaxial resonance isolator here because the coaxial junction isolator is currently more popular and offers better electrical performance for most applications.

REFERENCES

1. Howe, H., *Stripline Circuit Design* (Dedham, MA: Artech House, 1974).
2. Jensen, C., and J. Helsel, *Engineering Drawing and Design* (New York: McGraw-Hill, 1979).
3. Laverghetta, T. S., *Modern Microwave Measurements and Techniques* (Norwood, MA: Artech House, 1988).
4. Scott, A. W., *Cooling of Electronic Equipment* (New York: John Wiley and Sons, 1974).
5. Hutchings, J. L., J. R. Nortier, and D. A. T. Lambert, "Contour Program Smoothes Strip Discontinuities," *Microwaves and RF*, November 1987, pp. 129–138.
6. Bostick, G., G. F. Kinnetz, and T. W. Parker, "Designing High-Power Series Inductors and Shunt Capacitors," *Microwaves and RF*, October 1987, pp. 97–107.
7. *Reference Data for Radio Engineers* (Indianapolis: Howard W. Sams, 1977).
8. Military Specification MIL-W-85G, April 1976.

Chapter 8
Assembly and Testing

8.1 ASSEMBLY TECHNIQUES

After the circulator design phase is complete, we assemble the circulator and evaluate the design. As with most microwave and RF components, circulators do not always perform exactly according to the design specifications. We must test circulator performance, compare test results with desired performance parameters, and redesign all or parts of the circulator to improve certain aspects.

Circulator assembly differs from general electronic assembly work in several ways. First, circulators are very precisely machined. To maintain this precision, certain special assembly techniques must be used. Second, circulators use strong magnets that require care in handling and installation. Circulators, being RF components, differ substantially from dc and 60 Hz ac wiring with which most electronic assemblers are familiar.

Rather than present specific assembly techniques for each class of circulator, we will present general techniques, most of which apply to any given type of circulator. It will be obvious which techniques apply to the particular circulator class being assembled. For example, information on coaxial connector assembly obviously does not apply to waveguide units, and bonding ferrites to waveguide walls obviously does not apply to coaxial units.

A good rule to follow in circulator assembly is to trust no one. Check individual circulator components carefully before assembly. Dimensions are typically very important. If the ground-plane spacing or ferrite thickness of a coaxial junction circulator are incorrect, the result could be fractured ferrites (ferrites are not cheap). If the pin depth of a coaxial connector is incorrect, expensive test equipment could be ruined when the circulator is connected. If the ferrite slabs in a waveguide circulator are not properly secured, they could work loose, degrading electrical performance, possibly causing failure of a very expensive RF power source. Checking individual parts to drawings does not necessarily guarantee that they are correct; the drawings could be wrong. It is best to check parts against each other as well as against drawings.

Anyone involved in circulator assembly or solving problems related to circulator assembly should know how to use precision measuring tools. The most commonly used tools are the caliper and the micrometer. Calipers are usually capable of inside, outside, and depth measurements; micrometers usually measure only one parameter, depth or outside dimension. For most work with circulators, resolution to 0.001 inch is sufficient, but at times we need to resolve to 0.0001 inch. Reading dial calipers and micrometers to one thousandth of an inch is fairly straightforward so long as the operator realizes that calipers are not a monkey wrench and a micrometer is not a C-clamp. Reading either dial calipers or micrometers to greater precision or reading vernier calipers is more difficult.

We explain how to read a vernier scale with the aid of Figure 8.1. The vernier scale of a micrometer that is shown consists of eleven thin lines [1], scribed parallel to the sleeve long line, marked 0 through 10. When the thimble mark does not line up exactly with the sleeve long line, we examine the vernier scale to read the fraction of the thousandth increment on the thimble. One of the vernier lines will be aligned better with one of the thimble marks than all the others. The number corresponding to this vernier line is the number of ten-thousandths to be added to the thimble reading. The reading shown in Figure 8.1 is 0.1633 inch.

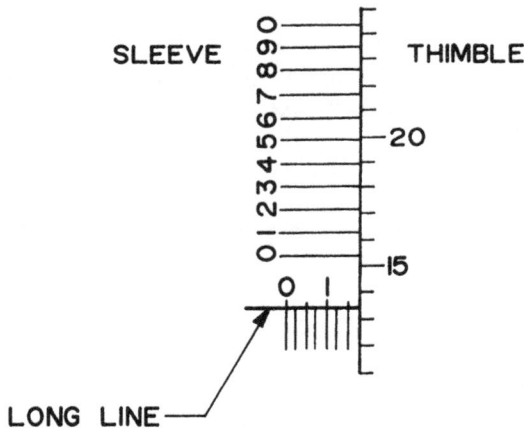

Fig. 8.1 Reading a vernier scale.

After the individual parts of the circulator are checked, they should be cleaned and dried if necessary. It is particularly important for the ferrites to be free of any coolant residue left over from the grinding operation. The ferrites can be washed in a volatile solvent and then baked to drive this solvent out of the pores and cracks.

Ferrite materials are easily confused with one another. Because the materials in bulk form are so similar in appearance, it is impossible to tell them apart by looking at them. Two relatively simple tests help us discriminate between materials if they become mixed up. We test the ferrite density and Curie temperature. Different ferrite materials are likely to have different densities or different Curie temperatures, so we can discriminate between the two if we know which is more dense or which has a higher Curie temperature.

The density is determined by weighing a sample of the material in question and dividing the weight by its volume. The volume can be either calculated from the ferrite dimensions or taken as the volume of water the sample displaces.

We compare Curie temperatures by sticking samples of the two materials under investigation on a magnet and placing the magnet on the ceiling of an oven. We turn on the oven (high temperature) and wait for the oven to warm up. When we hear a ferrite drop from the magnet, we know that the ferrite still on the magnet probably has a higher Curie temperature. Not all ferrites can be compared using this method because most ovens do not reach high enough temperatures to accommodate ferrites having high Curie temperatures. If the ferrite that falls breaks into a hundred pieces and is the only one of its type and the one you needed, you did something wrong.

The next step in the assembly procedure is to mount the ferrites. In some cases, they are metalized and soldered in place or bonded with conductive epoxy. We make an effort to eliminate air pockets between the ferrite and ground plane or waveguide wall. The quantity of solder or epoxy should be sufficient to form a good bond, without oozing onto the ferrite edges.

In coaxial junction circulators, the ferrites should be temporarily held in place (until final assembly, when compression will hold the ferrites) using silicone grease or petroleum jelly. A thin coating is all that is necessary. The grease tends to fill any air pockets that may be present on the ferrite surface, improving thermal conductivity and RF ground contact. When the ferrites are firmly held in place, it is not necessary to use assembly grease. The ferrites must be properly positioned, usually centered over a magnet well. In addition, the ferrites should be in registration with each other. Figure 8.2 illustrates ferrite centering. Ferrites that are not automatically centered by mechanical means usually need to be centered within a few thousandths of an inch to ensure good electrical performance. We offer no proof for this figure. Most people can see that something is off center a few thousandths on the scale of a typical circulator.

To improve electrical contact between the ferrite faces and ground planes in circulators that do not utilize ferrites bonded to the ground planes, we treat the

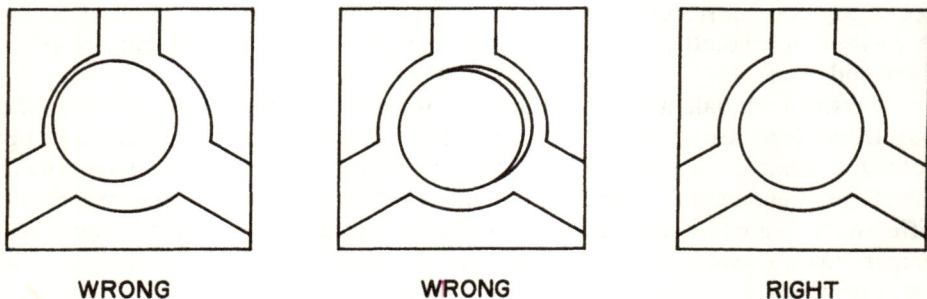

| WRONG | WRONG | RIGHT |

Fig. 8.2 Ferrite centering.

ferrites with aluminum foil. A thin coating of adhesive or grease is applied to the ferrite face. Then we spread a piece of aluminum foil on the ferrite face nearest the ground plane. The foil must be free of wrinkles and air pockets. The electrical contact is improved in many cases because the ground plane may not be perfectly smooth, but the foil bridges any valleys that exist in the ground plane surface. The foil does not significantly affect the ground-plane spacing because most foils are much less than 0.001 inch thick.

After the ferrites are installed, we proceed to install the connectors of a coaxial circulator. There are usually no other internal components of a waveguide circulator, with the possible exception of dielectric slabs. In the case of the lumped-element circulator, the ferrites are not installed until after the web-like center conductor has been wrapped around them. The coaxial connectors should be attached to the bottom half (the half with the threaded holes for holding the housing halves together) so that they will stay with this half when the top half or cover is removed. If the connectors and center conductor (which is usually soldered to the connectors) stay with the top half when the circulator is disassembled, it is more likely that the ferrites will move around. In addition, we usually want to look at the center conductor when the circulator is disassembled, as the reason for disassembly is for tuning purposes. The top cover or half is defined as the part with the clearance holes for the assembly screws because it makes more sense to work with the circulator when it is positioned with the screw heads on top for easy access. Figure 8.3 illustrates a partially assembled coaxial junction circulator.

Coaxial connectors with uncaptivated center pins are often used for circulators because they save the space normally occupied by the captivation mechanism. When the circulator is assembled, the center conductor is soldered to the connector center pin, providing the means of captivation.

Where the circulator center conductor provides the means of captivation, a

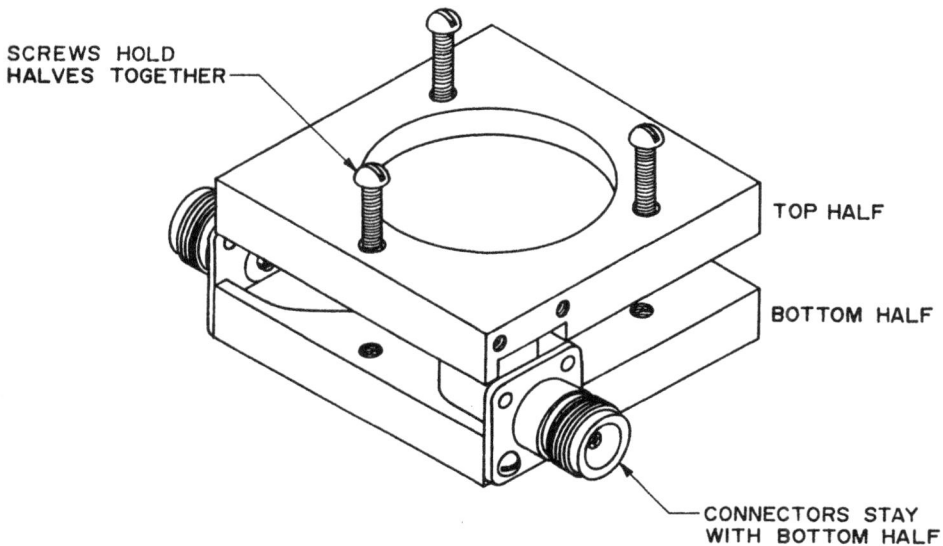

Fig. 8.3 Partially assembled coaxial junction circulator.

strong mechanical joint must be made between connector pin and center conductor. We make this joint by brazing, silver soldering, or conventional soldering, in order of decreasing strength [2]. Before and after we make the connection, we must be sure the connector pin depth is correct. We can do this most accurately using a connector gauge. We can also measure the pin alignment with conventional measuring tools. A simple method of holding the connector pins at the correct depth during the soldering process is to mate the connector with another connector that will hold the center pin in alignment and at the correct depth. Dimensions for connector pins are available from connector manufacturers and various reference books [3].

The next step in the circulator assembly process is temporarily closing the housing. We do not permanently close the circulator at this point because we may have to open the unit later to perform mechanical modifications for tuning. At low frequencies, it is not necessary to install all of the screws that secure the coaxial connectors. The circulator can be electrically tested with missing connector screws. At higher frequencies it is necessary to install most or all of the connector screws to avoid degradation of electrical performance due to poor grounding of the connectors.

Unless the magnets will be charged during circulator tuning, we need to start with oriented (magnetized, charged) magnets. The magnets can be either purchased already magnetized or magnetized just before installation. Several types of magnet chargers are available. One of these is the capacitive-discharge magnet-izer. This device utilizes capacitors that are charged to a specific voltage depend-ing on the magnetizing force required. To charge the magnet, the capacitors are connected to a low-resistance electromagnet. The magnet is placed in a magnetic circuit with the electromagnet so that the current pulse and resulting magnetic pulse charges the magnet. Another type of charger operates by rectifying the ac line voltage and applying the resulting dc pulses to the charging electromagnet. The amount of magnetizing force depends on which portion of the ac line half-cycle is applied to the electromagnet; this is done with silicon-controlled rec-tifiers.

After we charge the magnets, we thermally stabilize them if they are to be used in a circulator with a broad operating temperature range. Magnets experi-ence an irreversible loss of magnetic field when they are exposed to a temperature cycle [4]. We can thermally stabilize the magnets by cycling them several times between temperature extremes equal to or greater than the specified temperature range for the circulator. Where both operating and storage temperature ranges are specified, the storage range is usually more severe and should be used for cycling the magnets. The magnets should not be recharged after temperature cycling because this would defeat the purpose of thermal stabilization.

The pole pieces and magnets should be temporarily installed in the circula-tor. The magnets may have to be removed for adjustment by partial demagnetiza-tion unless equipment is available that will enable the magnets to be calibrated (adjusted) while they are installed in the magnetic circuit. It is best, especially with regard to circulators that are engineering development models, not to install the magnetic shields until the circulator is nearly completed. If the shields are not present, the magnetic field intensity can be experimentally adjusted by either holding other magnets near the magnets in the circulator or placing pieces of steel against the magnets to increase the magnetic field.

8.2 FINDING THE OPERATING POINT

After the preliminary assembly of the circulator is complete, we proceed to find the operating point. By this, we mean we adjust the magnetic field intensity until no further improvement in electrical performance can be obtained. The best elec-trical performance may not be obtained at the correct operating frequency of the circulator, and it may not meet the specifications. Nevertheless, magnetic adjust-ment is a good starting point for several reasons.

First, it is difficult for the beginner to design a magnetic circuit accurately so that no adjustments are required because there are empirical factors due to leak-age flux, as we discussed in Section 6.1. Second, magnetic adjustment is easy. To determine in which direction the adjustment must be made, we can simply hold a magnet near the circulator in either a bucking or aiding orientation. Bucking refers to the orientation that reduces the magnetic field supplied by the internal circulator magnetic circuit, and aiding is the orientation that increases the field. Magnetic adjustments made by demagnetizing or magnetizing the circulator magnets are not permanent—we can always recharge the magnet one more time. Electrical adjustments made by mechanical modifications are not so easily undone. Lastly, changes in magnetic field typically have a profound effect on circulator performance compared to small changes in impedance-matching circuit component values, which are not so easily changed in many cases.

We start by connecting the circulator to a calibrated circulator test setup. Many different test setups are used in the industry, ranging from antique signal generators and slotted lines to automatic vector network analyzers. A typical test installation was previously depicted in Figure 2.3.

Scalar measurement techniques (techniques that do not include phase measurements) can be used to test most circulators; however, phase information can be quite useful for determining how electrical adjustments should be made. For this reason, it is advisable to use test equipment that will provide complex transmission and reflection information, if at all possible. To test differential phase shift sections that will be part of differential phase shift circulators, the transmission phase shift information is a necessity.

The most important electrical parameter to monitor as we adjust the magnetic field is the insertion loss. The phase shift is also important for differential phase shifters. If the magnetic field is not correct, the insertion loss will be high over all or part of the circulator operating frequency range. Below resonance (in the magnetic domain, not the frequency domain), insufficient magnetic field leads to unsaturated ferrite and low-field losses. If the magnetic field intensity is set too high, resonance losses will occur. For resonance isolators, the magnetic field must be set at resonance. At resonance, the insertion loss of the isolator is at a minimum and the isolation (reverse loss) is maximal. Above resonance, insufficient magnetic field intensity leads to resonance losses. Excessive magnetic field reduces κ/μ (see Equation (5.48)), hence the splitting between the two counter-rotating modes, reducing the circulator bandwidth and thus the increasing insertion loss at the band edges. Figure 8.4 shows insertion-loss frequency-response curves for above- and below-resonance circulators with incorrect magnetic field intensities.

Our goal in finding the operating point is to adjust the magnetic field intensity until no further improvement in electrical performance can be obtained. Most circulators require only that we minimize the insertion loss over the operating frequency range. We do not need to make permanent changes to the magnetic circuit yet, because we will very likely have to readjust the magnetic bias later on, after we have performed some electrical adjustments.

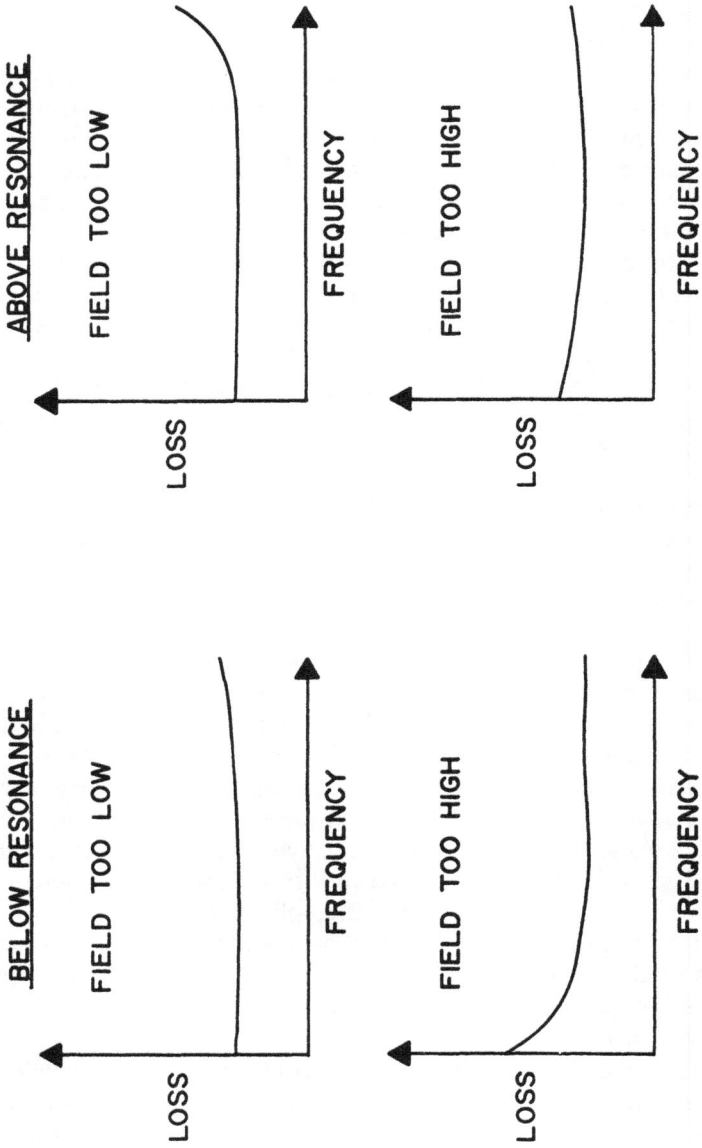

Fig. 8.4 Frequency-response curves for circulators with incorrect magnetic field.

Unless reference magnets or some other means of measuring magnetic field polarity are on hand, we will not know in which direction the circulator is circulating. It is important to remember that the magnetic field polarities for above- and below-resonance circulators are opposite for a given circulation direction. We must determine the direction of circulation before we can monitor the insertion loss. This is easily done by trial and error. We know from our magnetic circuit design computations approximately how much magnetic field will be needed to bias the ferrites. If we start with this level of magnetic field, the circulator will probably have higher insertion loss in one direction than in the other. The direction that has lower loss should be taken as the forward (insertion loss) direction. If we have specified a particular port as the circulator or isolator input, we have a fifty-fifty chance of installing the magnets correctly the first time. We will surely get it right the second time.

We need not be concerned if the insertion loss does not meet specification after preliminary magnetic adjustments are made. If the insertion loss is high because the VSWR is high (mismatch loss), we may be able to bring the insertion loss within specification by making some electrical adjustments. If the VSWR is within specification but the loss is too high, it may be that our operating point is not exactly right. We may have to change the magnetic field intensity and then electrically tune the circulator. Both of these situations and others will be discussed in more detail in Chapter 9.

We can make the temporary magnetic field adjustments described in this section by stacking up magnets of various sizes and shapes on top of the circulator magnets. Small magnets make small changes in the magnetic field and large magnets make large changes (assuming all the magnets are made of the same material and are charged to the same degree). The magnets can either aid or buck the circulator magnets. We can reduce the magnetic field by placing more pole pieces or spacers between the circulator magnets and pole pieces. The field can be increased by small amounts using pieces of steel or iron placed in contact with the magnets.

8.3 TAKING DATA

Because circulator design is typically an iterative process involving both prediction and experimentation, data collection and manipulation form a very important aspect of any circulator design algorithm. It would be absurd to synthesize a circulator on paper, build it, and install it in a system without ever checking the performance of the unit. We must be able to measure the electrical parameters of the circulator and record them for future use in an intelligent manner.

Several terms describe various qualities of measurement systems [5]. Perhaps the most important is accuracy. Accuracy is the deviation of a reading from a known input. We should know the accuracy of our test equipment, because the accuracy of the data and the quality of the circulator depend on it. Precision is the ability of an instrument to reproduce a certain reading with a given accuracy. Precision does not imply accuracy; an instrument may display five digits but have an accuracy of $\pm 1\%$. In this case, three of the five digits displayed may not be correct. It is of no help to record five digits if we are certain that three of them may be wrong unless we are only comparing measurements made using the same instrument. Engineers often speak of measurement uncertainty. The uncertainty of a measurement is the plus-minus range of the instrument accuracy.

Calibration of test equipment is important to ensure accuracy. Most facilities have calibration procedures that are used to maintain instrument accuracy. Instruments are checked against either a primary or secondary standard on a regular basis. A primary standard might be one at the National Bureau of Standards and a second standard could be an instrument similar to the one being calibrated but of known accuracy. Most calibrated test equipment has accuracy somehow traceable to the National Bureau of Standards.

Circulator isolation and insertion loss are normally measured in decibels. Reflection can be expressed in terms of voltage reflection coefficient, return loss, or VSWR. Equations (2.4) and (2.5) are useful for performing conversions between these units.

A bound notebook should be maintained to record circulator performance data. It will be necessary to look at the data when we decide how to make electrical and magnetic adjustments. Test data are often sent along with circulators if they are shipped. If anything unusual or significant is observed during the circulator development, this information could be used later on for other circulators, or could become part of a patent application.

Most modern microwave test equipment is capable of producing test reports of one kind or another. If such equipment is not available, the test results are recorded by hand. Whether the data are recorded in tabular or graphical format, enough points should be used to describe adequately the shape of the circulator frequency response. It is often helpful to take data over a broader range of frequencies than the circulator's operating frequency range. Broadband data may provide additional clues as to how to tune the circulator.

REFERENCES

1. Stockel, M. W., *Auto Service and Repair* (South Holland, IL: Goodheart-Willcox, 1969).
2. Jensen, C., and J. Helsel, *Engineering Drawing and Design* (New York: McGraw-Hill, 1979).
3. Laverghetta, T. S., *Modern Microwave Measurements and Techniques* (Norwood, MA: Artech House, 1988).
4. *Factors Affecting Magnet Stability* (Valparaiso, IN: Indiana General, 1978).
5. Holman, J. P., *Experimental Methods for Engineers* (New York: McGraw-Hill, 1984).

Chapter 9
Tuning

9.1 MAGNETIC ADJUSTMENT

In order to tune a circulator properly, especially a model that has not been built before, we must alternate between magnetic and electrical adjustments. The process usually begins with magnetic adjustment as described in Chapter 8. The reason for alternating between magnetic and electrical adjustments is that we want to see the results of each to determine whether an improvement in circulator performance has been made. In addition, it is difficult for one person to make both adjustments at the same time.

If we assume the dc magnetic field applied to the ferrites is uniform, as it should be if the pole pieces are of sufficient thickness and area, the only magnetic adjustment we need to make is to the magnitude of the field. If the circulator is to operate over a broad temperature range or have special magnetic shielding, there may be more tuning involved, but these cases are special. In this section we present information about how changes in the magnitude of the dc magnetic field affect circulator performance.

The electrical parameters with which we are concerned during tuning are the frequency, bandwidth, insertion loss, isolation, and VSWR. The isolation of a circulator is dependent on the return loss of the isolated port as described in Chapter 2, so we need to look only at the frequency, bandwidth, insertion loss, and VSWR or return loss at all ports.

Tuning a circulator is largely an impedance-matching problem. For a given reactive load, there exists a theoretical limitation on broadband impedance matching. Let us consider Figure 9.1. The lossless impedance-matching network is used to match the load, R_1 and C_1, to the source resistance, R_0, for maximum power transfer. Bode [1] showed what the impedance-matching limitations were, and Fano [2] later presented more general limitations on the matching of any load.

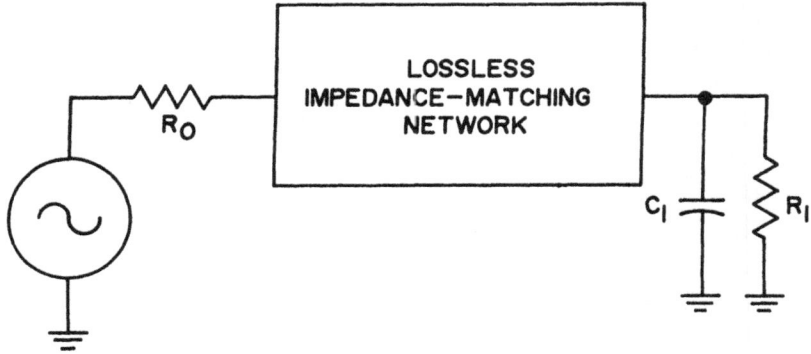

Fig. 9.1 Impedance matching of a reactive load.

The equivalent circuit of a circulator junction is typically a parallel-resonant circuit, so the load in Figure 9.1 nicely represents the circulator junction on one side of resonance. The capacitor C_1 could be replaced by an inductance and the same analysis used by duality.

The best possible impedance match is achieved when [3]

$$\int_0^\infty \ln \left| \frac{1}{\rho} \right| d\omega = \frac{\pi}{R_1 C_1} \qquad (9.1)$$

where ρ is the voltage reflection coefficient looking into the source side of the impedance-matching network. For the best impedance match in the circulator operating frequency range, we want the reflection coefficient to be equal to unity for all frequencies outside this range. Under this condition, we write

$$\int_{\omega_1}^{\omega_2} \ln \left| \frac{1}{\rho} \right| d\omega = \frac{\pi}{R_1 C_1} \qquad (9.2)$$

where ω_1 and ω_2 are the edges of the circulator frequency range. If ρ is constant across the frequency band, we can readily evaluate the integral in Equation (9.2) and write

$$\rho = \exp \left[\frac{-\pi}{(\omega_2 - \omega_1) R_1 C_1} \right] \qquad (9.3)$$

We can see from the preceding equations that we will achieve the best impedance match if the frequency response of the circulator has the shape of a bandpass filter: high VSWR outside the passband and low VSWR in the passband. We will have a lower passband VSWR if the bandwidth is more narrow or if the product $R_1 C_1$ is lower. For a broadband match, we want a low-resistance load with a low reactive component (low Q).

Our goal in tuning a circulator, then, is to make the passband only as broad as necessary, keep the Q as low as possible, and minimize the insertion loss. We will now consider what effects changes in magnetic field have on these parameters.

For the above-resonance mode, the splitting factor is approximately given by

$$\frac{\kappa}{\mu} = \frac{M_0 \omega}{\gamma H_{\text{dc}}(H_{\text{dc}} + M_0)} \tag{5.48}$$

From Equation (5.15), we know that the circulator bandwidth is proportional to the splitting factor. Hence, the bandwidth (by inspection of Equation (5.48)) is roughly inversely proportional to the square of the internal magnetic field:

$$\% \text{ BW} \propto \frac{1}{H_{\text{dc}}^2} \tag{9.4}$$

The bandwidth increases dramatically with decreases in magnetic field.

The calculation of insertion loss as a function of applied magnetic field is rather involved, but can be carried out as described in Section 5.1. We can more easily see the effect of changes in magnetic field by examining Figure 5.10. The insertion loss climbs rapidly as we decrease the magnetic field, approaching resonance.

Generally, the best magnetic adjustment for the above-resonance circulator is found as follows. We set the field low enough so that the insertion loss just meets specification at the highest frequency in the band, considering mismatch loss. This will give us the broadest possible bandwidth without sacrificing insertion loss, simplifying the impedance-matching process. Sometimes, for reasons of temperature stability, we choose to use a higher magnetic field intensity. This increases the degree of impedance-matching difficulty and, perhaps, circuit complexity.

Circulators exhibit mismatch losses due to both the input and output VSWRs. Mismatch loss is given by

$$\text{Mismatch loss} = -10 \log_{10}(1 - \rho^2), \quad \text{dB} \tag{9.5}$$

where ρ can be computed from VSWR using Equation (2.4). The two mismatch losses should be added together. The true circulator insertion loss, in a matched condition, is equal to the measured loss minus the sum of the input and output mismatch losses.

The reason circulators have mismatch losses at both input and output, rather than one mismatch loss as for most low-loss linear components, is that any power reflected from the output port is absorbed in the termination at the third port. In a reciprocal component, the power reflected from the output mismatch would be transmitted in the reverse direction back to the input (with low loss), where it would combine with the power reflected from the input to form a net reflected power that is probably less than the sum of the two reflected signals. This concept is illustrated in Figure 9.2.

RECIPROCAL COMPONENT   CIRCULATOR

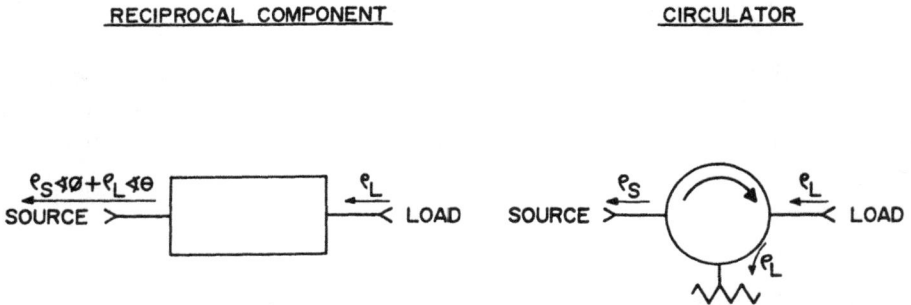

Fig. 9.2 Circulator mismatch losses.

For the below-resonance mode, the magnetic field adjustment is not as critical as that for the above-resonance mode. The splitting factor and bandwidth are much more dependent on the ferrite saturation magnetization, as we discussed in Chapter 5. An increase in magnetic field does, however, increase the bandwidth somewhat [4].

The magnetic field intensity must be sufficient to saturate the ferrite material to avoid low-field losses. It is unlikely that the field will ever be so strong that resonance losses occur, because this would require a field of very large magnitude at the frequencies where below-resonance circulators normally operate.

9.2 ELECTRICAL ADJUSTMENT

Another aspect of circulator tuning is electrical adjustment. The electrical adjustment of a circulator is primarily an impedance-matching problem. We can make changes in the impedance characteristics of stripline junction circulators by modifying the geometry of the portion of the center conductor between the ferrites, but there is no straightforward method of analyzing the effect of these geometrical modifications.

The first step in an electrical adjustment procedure is to evaluate the impedance of the port or ports to be matched. In order to measure properly the impedance of one port of a three-port circulator, the other two ports should be matched; this can be done with stub tuners or some other temporary means. As a rule of thumb, the two ports should be matched to achieve 10 dB return loss or more. Because a reciprocal three-port junction would ideally exhibit 10 dB return loss at each port with the other two ports terminated in matched loads, and only a nonreciprocal three-port junction can be perfectly matched [5], we know the circulator junction is nonreciprocal (certainly a desired quality) if we have 10 dB or more return loss.

After we have measured the complex impedance of the port to be matched, we need to de-embed the circulator junction or characteristic plane. This process will give us the impedance we wish to transform to the system-characteristic impedance (usually 50 Ω). The de-embedding process usually stops at the edge of the ferrite material, or at the characteristic plane in the special case of a waveguide junction circulator. Figure 9.3 illustrates the concept of de-embedding.

There are several ways we can perform the de-embedding. One is to use a computer analysis program. This is the fastest and easiest method if the software is available. Another method is to perform the necessary calculations by hand, which is quite tedious. A third method is to use the Smith chart, which is the method we will describe here.

Our impedance calculations fall into two classes: those for transmission lines and those for lumped elements (discrete inductors and capacitors). We will first consider transmission lines.

Having measured the complex impedance, we can proceed to find the impedance at a point inside the circulator, normally at the ferrite edge. Because we will be moving away from the generator, we need to begin with the conjugate of the measured impedance. This conjugate has the same real component, but the imaginary component has the opposite sign.

For simplicity, we normalize the Smith chart to the characteristic impedance of each transmission line through which we travel. The de-embedding process is best illustrated using an example. Our example is shown in Figure 9.4. The measured impedance is 63 +j10 Ω. The complex conjugate of this is 63 −j10 Ω. The first transmission line section in the circulator has a characteristic impedance of 37 Ω and an electrical length of 90 degrees. We normalize the Smith chart in Figure 9.5 to 37 Ω and plot 63 −j10 Ω on the chart; 63 −j10 is normalized as follows:

$$\frac{63 -j10}{37} = 1.70 -j0.27 \tag{9.6}$$

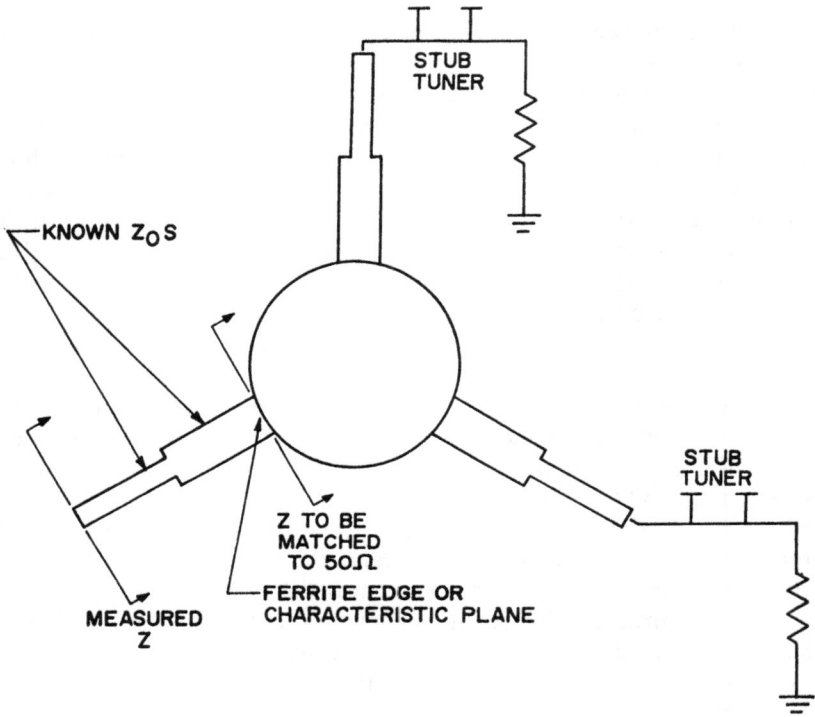

Fig. 9.3 Circulator junction de-embedding.

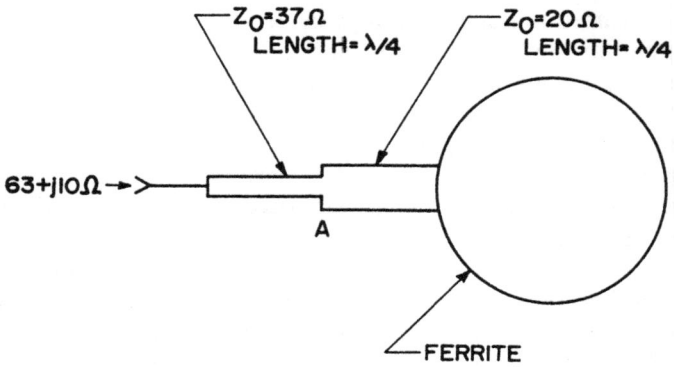

Fig. 9.4 Example of de-embedding.

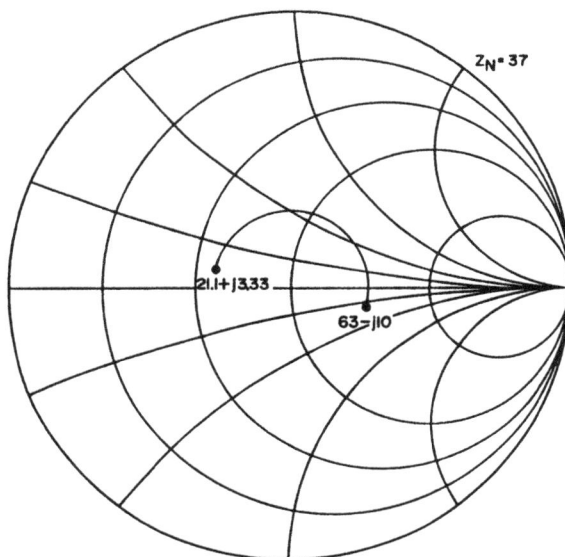

Fig. 9.5 First step of de-embedding.

We rotate this point through 180 degrees on the Smith chart (90 degrees transmission length) toward the load, counterclockwise. We arrive at the normalized impedance looking toward the source at point A in Figure 9.4. The impedance at this point is approximately $0.57 + j0.09 = 21.1 + j3.33 \ \Omega$.

We now renormalize a Smith chart to 20 Ω, the characteristic impedance of the transmission line section between point A and the ferrite. We plot $21.1 + j3.33$ Ω on the chart in Figure 9.6:

$$\frac{21.1 + j3.33}{20} = 1.06 + j0.167 \tag{9.7}$$

We again rotate the point 90 degrees toward the load and arrive at $0.92 - j0.13 = 18 - j2.6 \ \Omega$, approximately. This is the impedance looking toward the source. The impedance looking into the ferrite is the conjugate of this, $18 + j2.6 \ \Omega$.

From this impedance, we can proceed to design a matching circuit which will match this impedance to 50 Ω. In this example we only consider one frequency point. In a real situation, we need to consider a band of frequencies. The general procedure is the same, except for two things. First, the transmission lines

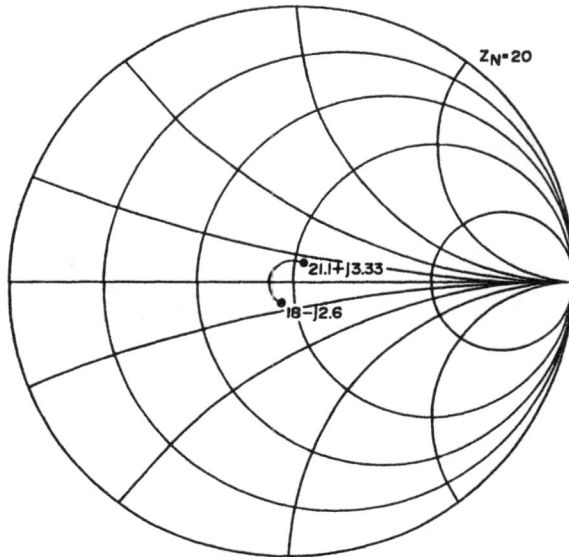

Fig. 9.6 Second step of de-embedding.

have different electrical lengths at different frequencies. Second, we need to consider how various components will affect the circulator frequency response. We do not want to use components that will reduce the bandwidth.

The impedance we want to match, $18 + j2.6 \, \Omega$, has a reactive component that can be handled in one of two ways [6]: by the absorption method or resonance method. In the absorption method we absorb the reactance into the matching network by modifying the original network slightly. The resonance method involves resonating the load reactance with an equal and opposite reactance.

For our example, we choose to use the absorption method because it will actually shorten the length of the transmission lines. This is not always the case, and it is usually undesirable to lengthen the transmission lines because the dielectric for the transmission lines may be ceramic, part of the ferrite assembly, and not easily enlarged. If we use the resonance method, the resonating section can be a shunt element, placed anywhere along the transmission line length without changing the dimensions of the dielectric.

To match $18 + j2.6 \, \Omega$ to $50 \, \Omega$, we begin with the existing $20 \, \Omega$ transmission line, normalizing the Smith chart in Figure 9.7 to $20 \, \Omega$. We plot the ferrite impedance ($0.92 + j0.13$). We rotate this point 120 degrees toward the generator to arrive at the real axis of the chart. Thus, we have transformed $18 + j2.6$ ($0.92 + j0.13$) Ω to $23.6 + j0$ ($1.18 + j0$) Ω using a 60 degree length of $20 \, \Omega$ transmission line. We can easily match $23.6 \, \Omega$ (real) to $50 \, \Omega$ using a 90 degree section of $34.4 \, \Omega$ line. Figure

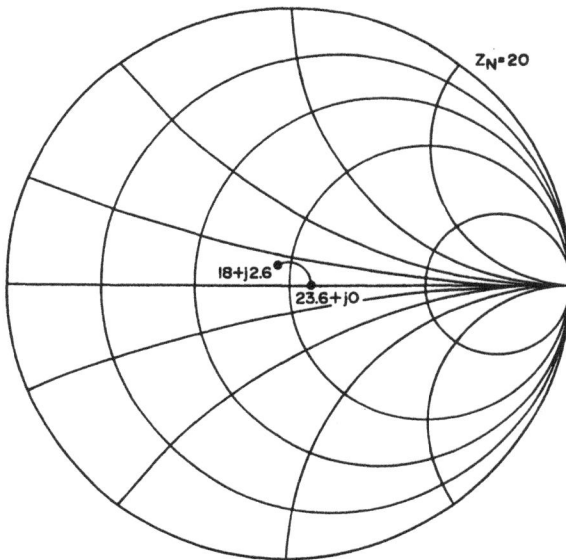

Fig. 9.7 First step of matching.

9.8 illustrates this last transformation.

After designing this new matching circuit, we modify the existing circuit accordingly. Insofar as the circulator is symmetrical, we perform the same modifications at each port.

We describe our de-embedding procedure for lumped-element matching circuits with the aid of Figure 9.9. We plot the complex conjugate of the measured impedance, $40 - j20$ Ω, on a Smith chart with normalized impedance and admittance coordinates [7] in Figure 9.10. The chart can be normalized to 50 Ω (0.02 S) or any other convenient impedance.

We next calculate the inductive susceptance of the 20 pF shunt capacitor at 200 MHz as 0.0251 S. We then normalize this susceptance to the normalization of the chart:

$$\frac{+j0.0251 \text{ S}}{0.02 \text{ S}} = 1.26 \tag{9.8}$$

We then increase the inductive susceptance of the source impedance already plotted on the chart by the amount of the inductive susceptance of the shunt capacitor by moving downward along a constant conductance circle 1.26 units. We arrive at an impedance of $0.25 - j0.43$ ($13 - j22$ Ω).

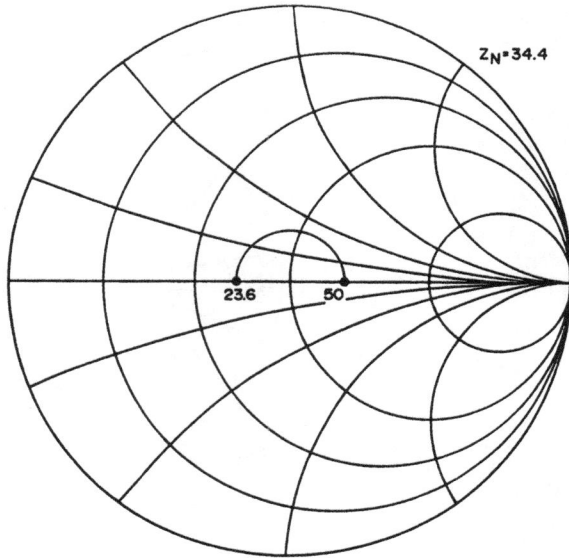

Fig. 9.8 Second step of matching.

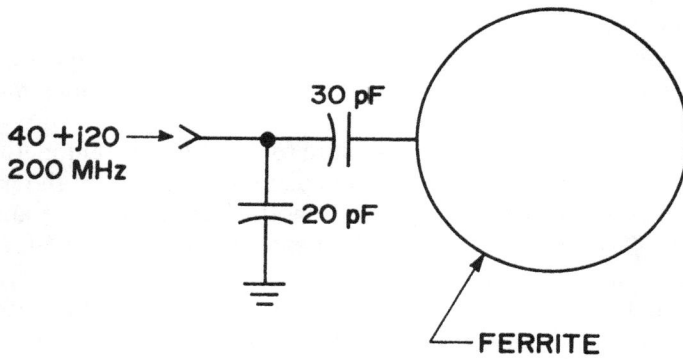

Fig. 9.9 Lumped-element de-embedding example.

Next, we calculate the reactance of the 30 pF series capacitor at 200 MHz as 26.5 Ω. This corresponds to a normalized reactance of

$$\frac{-j26.5}{50} = 0.531 \tag{9.9}$$

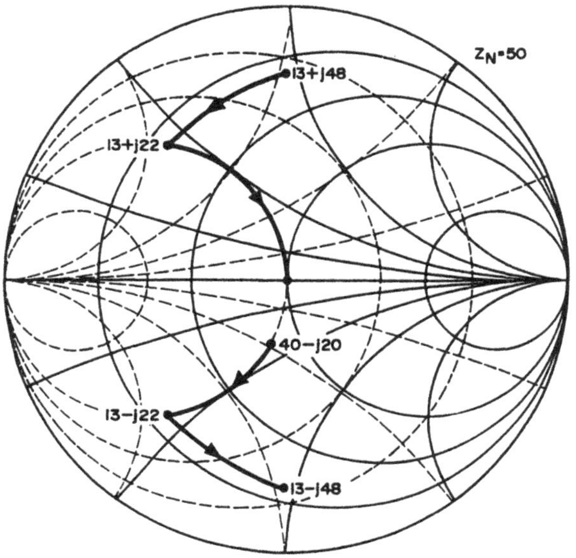

Fig. 9.10 Lumped-element de-embedding and matching.

We increase the capacitive reactance of the last point plotted by moving downward along a constant resistance circle 0.531 units. We arrive at an impedance of 0.25 −j0.96 (13 −j48.1 Ω). The conjugate of this value, 13 +j48.1 Ω, is the ferrite junction impedance.

By inspection of the Smith chart, we can see that the existing series capacitor brings the ferrite impedance to the 0.02 S conductance circle. If we simply adjust the value of the shunt capacitor, we will obtain a 50 Ω match. Starting with the ferrite impedance and adding the series capacitor in Figure 9.10, we arrive at an admittance of 1.0 −j1.76. This indicates that if we add an inductive susceptance of 1.76, we will be at the center of the chart. Thus, we calculate the value of shunt capacitance that will give us (1.76 × 0.02 S =) 0.0352 S. This value is 28 pF. Changing the shunt capacitor value to 28 pF will achieve a 50 Ω match as shown in Figure 9.10.

The same circuit modifications should be performed for all ports of a symmetrical circulator.

The effects of adding several different types of lumped components for impedance matching are shown in Figure 9.11. The effects on the circulator bandwidth should also be considered.

The material presented here is by no means complete in terms of impedance-matching techniques, but we have presented some basic concepts. There is often

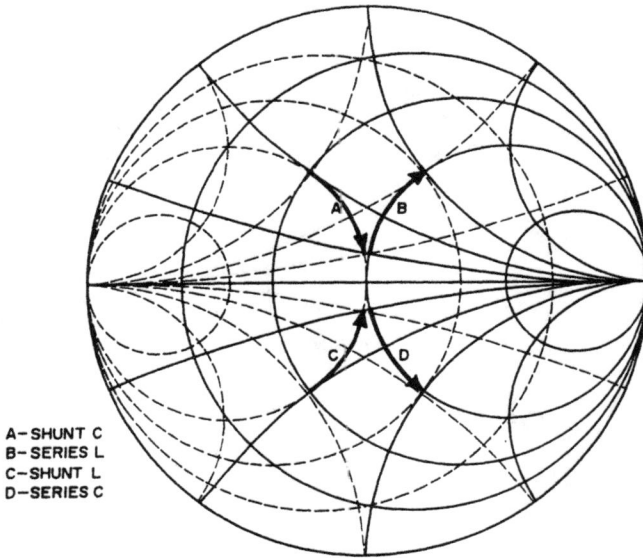

A—SHUNT C
B—SERIES L
C—SHUNT L
D—SERIES C

Fig. 9.11 Effects of lumped impedance-matching components.

more than one solution to an impedance-matching problem, and it takes experience and knowledge to arrive at the best one.

After an impedance match is achieved at each circulator port, the magnetic field can be adjusted again and the impedance-matching process repeated if there is a possibility of electrical performance improvement.

At microwave frequencies, it is easy to perform impedance matching by empirical means. If the inside of the circulator is accessible while RF is applied, we can probe inside the circulator with a metallic probe. A probe held in contact with the ground plane and in proximity with the center conductor will simulate shunt capacitance. If this capacitance makes the electrical performance of the circulator better, we can add shunt capacitance at this point. If the response worsens, we know to remove capacitance. Similarly, we can use a magnet and a piece of steel to probe a waveguide unit. The steel is placed inside the waveguide, and the magnet, outside the waveguide, is used to slide the steel around until good electrical performance is obtained. Then we either add a tuning screw at this location or dimple the guide.

REFERENCES

1. Bode, H. W., *Network Analysis and Feedback Amplifier Design* (New York: D. Van Nostrand, 1945).

2.	Fano, R. M., "Theoretical Limitations on the Broadband Matching of Arbitrary Impedances," *Journal of the Franklin Institute*, January-February 1950.

3.	Matthaei, G. L., L. Young, and E. M. T. Jones, *Microwave Filters, Impedance-Matching Networks, and Coupling Structures* (Dedham, MA: Artech House, 1980).

4.	Simon, J. W., "Broadband Strip-Transmission Line Y-Junction Circulators," *IEEE Transactions on Microwave Theory and Techniques,* May 1965, p. 339.

5.	Simon, J. W., "Broadband Strip-Transmission Line Y-Junction Circulators," *IEEE Transactions on Microwave Theory and Techniques,* May 1965, p. 344.

6.	Besser, L., *RF Circuit Fundamentals, Part 1* (Palo Alto, CA: Besser Associates, 1986).

7.	Bowick, C., *RF Circuit Design* (Indianapolis: Howard W. Sams, 1982).

List of Symbols

a = Waveguide width, coaxial dielectric inner diameter

b = Magnetic induction, ground-plane spacing, waveguide height, coaxial dielectric outer diameter

c = Velocity of light = 3×10^8 m/s

d = Ferrite thickness, wire diameter

e = Unit electron charge = 1.602×10^{-19} coulomb

f = Frequency

f_0 = Center frequency

f_1 = Lower frequency limit

f_2 = Upper frequency limit

f' = Scaled frequency

g = Landé splitting factor (g factor)

g_{eff} = Effective g factor

h = Magnetic-field intensity, Planck's constant = 6.625×10^{-34} J/s

i = Current in electromagnet

k = Wave number, thermal conductivity

m = Electron magnetic moment, magnetization

m_0 = Electron mass = 9.11×10^{-31} kg

n = Number of heat sink fins

r = Radius, circulator input VSWR

s = Resonator spacing

t = Time, stripline thickness, dielectric thickness

w = Heat sink fin spacing

x = Position of ferrite slab in waveguide, coordinate axis

y = Coordinate axis

z = Heat sink fin height, coordinate axis

$\hat{z}$ = Unit vector in z direction

A = Area, ferrite triangle altitude, altitude derating factor

A_g = Area of air gap
A_i = Area of iron or steel
A_m = Magnet area
A_{eff} = Effective ferrite triangle altitude
B = Circulator junction susceptance
B_d = Magnet induction at operating point
B_g = Magnetic induction in air gap
B_i = Magnetic induction in iron or steel
B' = Circulator junction susceptance slope parameter
BW = Bandwidth
C = Capacitance
C_f' = Fringing capacitance
D = Inductor coil diameter, determinant
F = Applied force, frequency, flow rate of water or air
G_r = Circulator junction conductance
H = Magnetic field intensity
H_a = Anisotropy magnetic field intensity
H_d = Magnet field strength at operating point
H_{dc} = Internal dc magnetic field intensity
H_{dc}' = Scaled dc magnetic field intensity
H_{ext} = Applied external dc magnetic field intensity
H_{RF} = RF magnetic field intensity
I = Isolation
I_{RF} = RF current
L = Angular momentum, length, inductance
L_g = Length of air gap
L_m = Magnet length
L_0 = Inductance of lumped-element circulator coil in air
$L_\pm$ = Inductance of lumped-element circulator coil for two modes
M = Total ferrite magnetization, integer
M_0 = Ferrite magnetization
M_0' = Scaled ferrite magnetization
N = External torque, integer, number of turns of wire in a coil, demagnetizing factors
P = Power
P_{in} = Average input power
Q = Power dissipated in circulator junction
Q_L = Loaded Q of circulator junction
Q_U = Unloaded Q of circulator junction
R = Ferrite radius, electromagnet winding resistance, resistance
R_g = Reluctance of air gap
R_i = Reluctance of iron or steel path

R' = Scaled disk radius
T = Temperature rise or drop
$\mathbf{T}$ = Generalized tensor
T_c = Curie temperature
V = Voltage
V_{dw} = Dielectric withstanding voltage
W = Stripline width, energy required to magnetize ferrite
Y_r = Circulator junction admittance
Y_0 = Characteristic admittance
Y_{eff} = Effective circulator junction admittance
Z_t = Transformer impedance
Z_0 = Characteristic impedance
α = Damping factor, attenuation constant
β = Phase constant, wave number
$\beta_{\pm}$ = Phase constants for two modes
γ = Gyromagnetic ratio = 2.8 MHz/Oe
γ_{eff} = Effective gyromagnetic ratio
δ = Dielectric loss angle
ε = Permittivity
ε_0 = Free-space permittivity = 8.85×10^{-12} F/m
θ = Angle
κ = Component of susceptibility and permeability tensors
κ' = Real component of κ
κ'' = Imaginary component of κ
λ = Wavelength
λ_c = Cutoff wavelength
μ = Permeability (scalar)
$\boldsymbol{\mu}$ = Permeability (complex)
$\underline{\boldsymbol{\mu}}$ = Permeability (tensor)
μ_z = Permeability in direction of applied dc magnetic field
μ' = Real component of μ
μ'' = Imaginary component of μ
μ_0 = Free-space permeability = $4\pi \times 10^{-7}$ H/m
μ_{eff} = Effective permeability
π = Archimedes' constant = 3.1415927
ρ = Voltage reflection coefficient
ρ_c = Circulator voltage reflection coefficient
ρ_L = Load voltage reflection coefficient
ρ_s = Source voltage reflection coefficient
σ = Electrical conductivity
ϕ = Angle, Faraday rotation, magnetic flux
χ = Susceptibility (complex)

χ = Susceptibility (tensor)
χ' = Real component of χ
χ'' = Imaginary component of χ
ψ = Stripline angle
ω = Precession radian frequency, microwave radian frequency
ω_0 = Ferrimagnetic resonance radian frequency
$\omega_{\pm}$ = Resonant radian frequencies of two modes
Γ = Propagation constant
$\Gamma_{\pm}$ = Propagation constants for two modes
Δd = Ferrite thickness correction factor
ΔH_k = Spin-wave line width
ΔH = Resonance line width
$4\pi M$ = Ferrite magnetization
$4\pi M_s$ = Ferrite saturation magnetization

Index

Above resonance
  definition, 34
  circulator design procedure, 83
Absorption method of matching, 180
Absorptive
  elements in dummy loads, 124
  waveguide elements, 124
Accuracy of measurement system, 177
Admittance, circulator junction, 83
Alloy, temperature compensation, 79
Altitude derating factor, 33
Aluminum substitution, 67
Amplifier
  buffer, 51
  negative resistance, 61
  stability, 51
Andradite, 5, 67
Angle units, 1
Angular momentum, 5
Anisotropy, 73
  field, 73
Antiferromagnetism, 7
Area units, 1
Assumptions, circulator design, 83
Attenuation constant, 20, 22
Auld, B.A., 98
Average power capacity
  cables, 33
  connectors, 33
  junction circulators, 37

Bandwidth
  circulator, 35
    below-resonance, 90
    differential phase shift, 42

equation, 92
  junction, 39
  lumped-element, 41
  waveguide junction, 93
  effect of decoupling on, 93
  expression for, 88
  isolator, 44
    field-displacement, 44
    resonance, 46, 121
  relation to magnetic field, 175
Bell Labs, 5
Below resonance, definition, 39
Bessel function, 88
BH characteristics of metals, 137
Bicycle wheel analogy, 9
Bode, H.W., 173
Bohr magneton, 12
Bonds, ionic, 5
Bosma, H., 87
Breakdown power, calculation of, 107

Calibration of test equipment, 172
Capacitors
  lumped-element circulator, 113–114
  procedure for design, 159–160
Center cavity size, 153
Center conductor, 87
  circulator, 39
  geometries, 99, 107
Ceramic dielectrics, 80
CGS system of units, 1
Characteristic impedance
  stripline, 105
  transformer, 93
Characteristic plane, 97

location, 97
Characteristics
  garnets, 67
  spinels, 67
Channel widths, 155
Circular polarization, waveguide, 118
Circulation
  directions of, 83
  perfect, 84
Circulator
  below-resonance, 90
  definition, 26
  failures, 25
  faraday-rotation, 13
  frequency range, 27
  parameters, 25
  octave-bandwidth, 98
  stripline, 87, 99
  synthesis procedure, 100
  waveguide junction, 93, 96, 111
Clarricoats, P.J.B., 117
Cleaning
  circulator parts, 164
  ferrites, 68
Coaxial circulator construction, 37
Coercive force, magnet, 78
Compression of ferrites, 149
Comstock, R.L., 87
Conductance, circulator junction, 102, 106
Conduction cooling, 153
Conductivity, ferrite, 7
Connectors, coaxial, 163
Construction, isolator
  field-displacement, 44
  resonance, 46
Contact, ferrite–ground plane, 148
Conversion
  factors, 1
  VSWR–reflection coefficient, 28
Cooling
  circulator, 40
    junction, 40
    methods, 152
  conduction, 153
  forced-air, 152
  natural-convection, 153
  water, 152
Coordination, ion site, 6
Critical power, circulator, 35
Crystal

ionic, 5
  plane, 6
  structure, 5
Curie temperature, 36, 70
  measurement of, 72
Cut-off wavelength, stripline, 103

Damping factor, 16
Data, recording of, 172
Davies, J.B., 87
De-embedding
  circulator junction, 177
  lumped-element procedure, 177
Decoupling, 109
Degenerate PARAMP, 59
Demagnetization curve, 134
Demagnetizing factors, 13, 133
Dielectric
  ceramic, 80
  classification, 79
  constant, 79
    ferrite, 80
    transformer, 99
  gases, 79
  loss tangent, ferrite, 70
  material, resonance isolator, 123
  silicone, 79
  withstanding voltage, 33
Differential phase shifts
  calculation of, 119
  implementation of, 117
Dimensional tolerances, ferrite, 69
Dimensions
  importance of, 163
  stripline center conductor, 87
Diode
  avalanche, 61
  BARITT, 61
  IMPATT, 61
  performance, 62
  TRAPATT, 61
Diplexer, 58
Disk radius, ferrite, 96
Dissimilar metals, 157
Doping, rare-earth, 76
Dunn, V.E., 114
Duplexer, ferrite, 53
Dyad, vector, 15

Eddy-current losses, ferrite, 9

Effective gyromagnetic ratio, 70
Electromagnet, 129–130
Electromagnetic shielding, 27
Electron spin, 9, 10
Ellipsoid model, 135
Elliptical polarization, 19, 21, 22
EMI shielding, 36
Energy
 product of magnet, 78
 required to magnetize ferrite, 129
English system of units, 1
Environmental factors, 36
Equivalent circuit, circulator, 94

Face-centered-cubic lattice, 6
Factors, conversion, 1
Fano, R.M., 173
Faraday effect, 13
Faraday, M., 13
Faraday rotation, 13, 21
Fay, C.E., 87
FCC lattice, 6
Ferrimagnetic
 materials, 4
 resonance, 9
 resonance absorption, 20
Ferrimagnetism, 4, 67
Ferrite, 4
 classes of, 5
 crystal structure, 5
 curie temperature comparison, 165
 density measurement, 165
 disk conversion to triangle, 97
 disk diameters, 89
  dimensions, 115
  radius, 88–89, 93, 98, 102, 106, 112
  thickness, 93, 96, 103
 garnet, 5
 hexagonal, 5
 hysteresis loop, 129
 magnetization, symbol for, 4
 material, 13
 mounting, 165
 polycrystalline, 5
 positioning, 165
 retentivity, 129
 shape, 84
 slab dimensions, 119
  position, 121–122
 sphere as test sample, 70

 spinel, 6
 temperature compensating, 79
 tests, to discriminate, 165
 thickness, 106, 112, 119
 triangle correction factor, 97
Ferromagnetism, 7
Field patterns, triangular, 86
Filling factor, 109
Finishing, 156
Firing of ferrites, 68
Flux density in shield material, 139
Forced-air cooling, 152
Forming of ferrite shapes, 68
Free magnetic poles, 133
Frequency
 limits
  field-displacement isolator, 44
  junction circulator, 37
  lumped-element circulator, 41
 range, 27
 resonance, triangular resonator, 98
 sensitivity, phase shifter, 63
 splitting, 91

G-effective value, 70
G factor, 12
Gadolinium, 67
Gain, negative resistance amplifier, 61
Garnets, 5, 7, 67
Geometry, circulator, 117
Gilbert equation of motion, 16
Grain size, ferrite, 69
Green, J.J., 87
Green's function, 88
Griffiths, 12
Ground-plane spacing, 98, 108
Ground planes, circulator, 37
Gyromagnetic ratio, 4, 12, 70
Gyroscope system, 10
Gyroscopic effect, 9

Harmonic generation, 35, 42
Helszajn, J., 87, 93–94, 97
Hexagonal ferrite, 5
Hilpert, 4
Holmium, 67
Homogeneity of ferrites, 68
Hoshino, N., 113
Humidity, 36
Hund's rule, 5

Hysteresis loop, ferrite, 129

Image plane, 97
Impedance, 83
  circulator input, 90
  matching, 98
    circuit design, 180
    limitations on, 173
    techniques, lumped-element, 107
Inductance
  coil, 114
  stripline, 116
Inductor design procedure, 159
Insertion loss, 27
  approximation of, 92
  calculation of, 92
  differential-phase-shift circulator, 117
  field-displacement isolator, 44
  junction circulator, 37
  lumped-element circulator, 41
  multiplexer, 57
  resonance isolator, 46
Insulation, strip, 116
Intermodulation products, 42
International system of units, 1
Ion
  metallic, 6–7
  oxygen, 6–7
Isolation, 27
  differential-phase-shift circulator, 117
  multiplexer, 58
  PARAMP, 59
  required, 28, 30
  resonance isolator, 46
  transmit-receive, 52
Isolator, definition of, 26

James, D.S., 97
Jet casting, 77
Junction
  circulator propagation constant, 21
  radius, microstrip circulator, 83
Kittel, 12
Kittel's equation, 134
Konishi, Y., 113

Laminates, printed circuit, 80
Landau and Lifshitz precession, 11
Landé splitting factor, 12
Lattice, crystal, 5
Leakage flux, 139

Length units, 1
Light, plane-polarized, 13
Limiter
  cavity, 54
  comb, 54
  ferrite, 53–54
  orthogonal stripline, 53
  subsidiary resonance, 54
Line width, resonance, 70, 75
Linear vector operator, 15
Loaded Q
  approximation of, 91
  calculation of, 91, 102
  circulator junction, 92
Local oscillator, 52
Lodestone, 4
Longitudinal field
  propagation constant, 18
Low-field loss, 73
Low-noise amplifiers, 52
Lumped components
  effects on bandwidth, 183
Lumped-constant construction, 41
Lumped-element
  circulator design procedure, 112
  matching techniques, 107

Machining
  ferrite, 68
  quality, waveguide circulators, 160
Magnet
  alnico, 77
  ceramic, 77
  chargers, 167
  coercive force, 78
  dimensions, calculation of, 135
  energy product, 78
  NdFeB, 78
  operating point, 138
  permanent, 131
  size, 132
  thermal stabilization, 168
  wells, 157
Magnetic
  bias, 87
  circuit, typical, 131
  field, 11
    adjustment, 168–169
    critical microwave, 22
    dc, 11–12

external, 13, 137
intensity, units, 1
requirements, lumped-element circulators, 112
RF, 12, 113
selection, 101
flux density, shield material, 139
induction, units, 1
losses
estimation of, 76
ferrite, 76
moment, 5
atomic, 7
electron, 12
ferrite ion, 5
net, 5
operating point, 73
junction circulator, 37
lumped-element circulator, 41, 115
microstrip circulator, 112
resonance isolator, 46, 123
operating regions, 33
below-resonance, 93
poles, free, 133
returns, 132
shielding, 36, 132, 139
shunts, 132
temperature compensating material, 140
data for, 140
Magnetite, 4
Magnetization, ferrite, symbol for, 4
Masers, 62
Matching, impedance
circuit design, 174
limitations on, 173
structures, 99
Materials, ferrimagnetic, 4
Matrix form of tensor, 15
Measurement system
accuracy, 171
precision, 172
qualities of, 171
Measurement tools, precision, 164
Metal, divalent, 6
Microstrip
circulator construction, 37
circulator design procedure, 111
line impedance, 112
Mismatch loss, 175
MKS system of units, 1
Mode, dominant, triangular resonator, 97

matching, analysis technique, 98
splitting
definition, 91
variation with frequency, 90
suppression, circular modes, 154
Modes
counter-rotating, 105
higher-order, 103, 109
of operation, 75
rotating, analysis, 83
National Bureau of Standards, 172
Natural-convection cooling, 153
Néel, L., 5
Nickel-steel alloy temperature
compensation material, 79
Noise figure, PARAMPs, 61
Nondegenerate PARAMP, 59
Nonlinearity, circulator, 35
Nonreciprocal properties, ferrite, 13
Notebooks, 172

Octants, 6
Operating point, magnet, 138
Operation of differential phase
shift circulators, 42
Orbital
atomic, 5
interactions, 5
motion, electron, 5
Outer conductor, coax termination, 125

Packaging schemes, stripline, 145
Paramagnetism, 5
Parameters, circulator, 25
Particle size, ferrite, 68
Peak power capacity, 69
coaxial connector, 31
junction circulator, 37, 40
lumped-element circulator, 41
Peak power mechanism for breakdown, 33
Percentage bandwidth, circulator, 35
Performance
differential-phase-shift circulator, 42
field-displacement isolator, 44
junction circulator, 37, 40
lumped-element circulator, 41
resonance isolator, 46, 121
waveguide junction circulator, 37
Permanent magnets, 131
Permeability
effective, 88, 101

below-resonance, 106
    simplified expression for, 89
magnetized ferrite, 13
measurements, 99
partly magnetized ferrite, 99
tensor, 15
unmagnetized ferrite, 100
z-axis, 100
Permittivity, symbol for, 4
Phase
    constant, 20
    shifter
        dual-mode, 63
        Fox, 63
        reflective, 63
        Reggia-Spencer, 63
        TEM, 63
        toroidal waveguide, 63
    term, of propagation constant, 18
Phillips Gloeilampenfabriken Labs, 4
Polder, 15
    tensor, 16
        derivation of, 16
Pole piece, 131
    designs, good and bad, 142
    discussion about, 141
    material for, 141
Polycrystal, 5
Polycrystalline ferrite, 68
Polyiron, 124
Power dissipation
    circulator junction, 151
    in ferrites, 109
Power handling (capacity), 25
    average, concerns, 107
    CW, waveguide, 32
    differential-phase-shift circulator, 42
    field-displacement isolator, 44
    junction circulator, 39–40
    lumped-element circulator, 41
    resonance isolator, 46, 49
    stripline, 153
Power level, PARAMPs, 60
Power limit
    average power, 33
    peak power, 32
Power threshold, ferrite, 31, 69, 76
Precession, 10, 11, 12
Precision
    measurement system, 171

measuring tools, 164
Presintering, 68
Printed circuit laminates, 80
Probing, empirical matching method, 184
Propagation
    constant, 18, 19
    in circulator junction, 83
    in resonance isolators, 46
Pulling, oscillator, 51
Pulse
    repetition rate, effect on waveguide power
        handling, 32
    width, effect on waveguide power handling,
        32
Pump source, PARAMP, 59

Q, loaded
    approximation of, 92, 106
    calculation of, 92
    circulator junction, 91
Q, unloaded, circulator junction, 92
Quantum numbers, 5
Quarter-wavelength impedance transformers,
    93, 95, 99, 106–107
Rare earth, 7
Rare-earth doping, 76
    of ferrites, 76
Ratio, gyromagnetic, 4
Raw materials, ferrite, 68
Receiver applications of isolators, 52
Reflection coefficient, 28
Reliability, circulator, 36
Reluctance, 137
Resistive elements, field-displacement
    isolator, 44
Resistors, thin-film, 124
Resonance
    absorption, 20, 48
    ferrimagnetic, 9
    frequency, triangular resonator, 97
    linewidth, 16, 70
        test for, 70
    operating region, 34
    subsidiary, 22
Resonator
    PARAMP, 59
Retentivity, ferrite, 129
Return loss, 28
Returns, magnetic, 132
Rexolite, 79

RF load parameters, 125
RFI shielding, 36
Roberts, R.W., 114

Sandy, F., 87
Saturation magnetization
    measurement of, 70
    selection of, 70
        above-resonance, 73
        below-resonance, 73
    symbol for, 4
Scaling, frequency, 111
Scattering matrix, 98
Schloemann, 100
Sealing,
    hermetic, 157
Separation of ports, PARAMP, 60
Series-resonant circuit
    lumped-element circulator, 113
Shape, resonance isolator, 46
Shells, atomic, 5
Shield
    design procedure, 140
    magnetic, 132
Shielding
    electromagnetic, 27
    magnetic, 27
        purposes of, 139
Shock, 27
    effect on lumped-element circulators, 41
Shunts, magnetic, 132
SI system of units, 1
Silicone dielectric materials, 79
Simon, J.W., 87
Single-crystal ferrite, 68
Sites
    ion, 6
    octahedral, 6
    tetrahedral, 6
Size
    circulators, 27
    differential-phase-shift circulators, 42
    junction circulators, 37
Smith chart, 179
Snoek, J.L., 4
Solid-state devices, 51
Spin
    axes, electron, 11
    waves, 18, 35
Spin-wave line width, 69

    measurement of, 72
Spinel, 5, 67
    ferrimagnetic properties of, 7
Stability
    amplifier, 51
    temperature, of ferrite, 75
Standards, primary and secondary, 172
Standing-wave pattern, 84
    ferrite, 84
    rotation of, 85
Strip width, lumped-element circulator, 113
Stripline, 103
    characteristic impedance of, 105
    circuit construction
        bonded-stripline, 148
        bonded-substrate, 145
        box-and-cover, 147
        channeled-plate, 147
        flat-plate, 145
    circuit packaging techniques, 145
    circulator construction, 37
    construction techniques, 148
    power-handling capacity, 151
    width
        calculation of, 105
        determination of, 105
        effect on coupling, 103
        restrictions on, 105
Subshell, atomic, 6
Subsidiary resonance, 22
Substrate thickness, microstrip circulator, 112
Surface finish, ferrite, 69
Susceptibility
    slope parameter, 95
    tensor, 17
Switch
    ferrite, 53
    reciprocal, 54
Symbol
    ferrite magnetization, 4
    permittivity, 4
    saturation magnetization, 4
Tan, F.C., 96
Teflon (PTFE), 80
Temperature
    Curie, 70
        ferrite material, 36
        measurement of, 72
    drop, across ferrite, 151
    effects on magnets, reversible, 78

ferrite surface, reduction of, 109
material
    data for, 140
    magnetic, 140
operating
    circulator, 26
    lumped-element circulator, 41
    magnet, 77
stability of ferrite, 78
storage, circulator, 36
Temperature compensation
using capacitors, 141
Tensor
definition, 15
permeability, significance of, 16
polder permeability, 16
Test
data, recording of, 172
set-up, typical, 29
Thermal
conductivity
    of selected materials, 151
    units, 1
performance, 109
stabilization of magnets, 168
Third-order intercept point, circulator, 35
Transformer, quarter-wavelength, 93, 95, 99, 106, 107
Transition
coaxial to stripline, 110
compensation method
    contour, 156
    triangular, 156
equivalent circuits, 156
models, 156
step, stripline, 156
Transverse field propagation constant, 20
Triangular resonator field patterns, 97
Tubes, microwave, 51

Uncertainty of measurements, 172
Units
angle, 1
area, 1
CGS system of, 1
English system of, 1
International system of, 1
length, 1
magnetic field intensity, 1
magnetic induction, 1

MKS system of, 1
propagation constant, 20
SI system of, 1
thermal conductivity, 1
used in this book, 1
Unloaded Q, circulator junction, 92
Vacuum tubes, microwave, 51
Vibrating sample method
saturation magnetization measurement, 70
Vibration, 36
effect on lumped-constant circulators, 41
Voltage breakdown, 107
calculation of, 107
VSWR, 27, 31
circulator, 92
junction circulator, 39
lumped-element circulator, 41

Water cooling, 152
Water-pipe analogy of circulator, 26
Wave
components, counter-rotating, 83
number, 88
Waveguide junction circulators, 37, 93, 96, 111
Web thickness, 150
Weight
circulator, 36
differential-phase-shift circulator, 42
junction circulator, 37, 40

YIG (yttrium iron garnet), 67
Yttrium, 7
Yttrium iron garnet (YIG), 67